Purple Plague in Microelectronics

This book is a comprehensive overview of the discoveries in composition, growth and control of gold-aluminum intermetallic compounds in microelectronic interconnects known as purple plague. It focuses on the history of detection, theory, and understanding of the gold-aluminum interface reliability issues.

Purple Plague in Microelectronics presents research on the phenomenon of purple plague from the earliest days to the present. The authors explain intricate scientific aspects to understand cause and effect of this reliability issue in microelectronics in a way that ensures the content is understandable to readers of various educational levels, novice to expert, while satiating the needs of the engineering personnel closely associated with work related to the gold-aluminum interfacial integrity. The discovery of void formation leading to porosity in the purple phase and the resulting joint failure is discussed, along with detailed analyses of the various phases, metallurgy characteristics, and diffusion phenomenon at the gold-aluminum interface. They provide detailed drawings, graphs, charts, models, and tables, which allow easy comprehension of concepts and phenomenon through visual aids. The book explains various methods for evaluating the quality and reliability of wire bond joints, highlighting the relative advantages of each approach. Techniques such as shear testing and bond resistance testing are discussed in detail. It also examines the role of ingredients used in packaging material and the gold wire itself, supported by relevant mathematical models.

Intended for electronic assembly engineers, design engineers, process development engineers, material scientists, electronic industry supervisors and managers, this book is an excellent reference for anyone involved in defect control and reliability in gold wire bonding.

Purple Plague in Microelectronics

Syed Sajid Ahmad and Richard C. Blish II

CRC Press
Taylor & Francis Group
Boca Raton London New York

CRC Press is an imprint of the
Taylor & Francis Group, an **informa** business

Designed cover image: Syed Sajid Ahmad

First edition published 2026
by CRC Press
2385 NW Executive Center Drive, Suite 320, Boca Raton FL 33431

and by CRC Press
4 Park Square, Milton Park, Abingdon, Oxon, OX14 4RN

CRC Press is an imprint of Taylor & Francis Group, LLC

ISBN: 9781041024248 (hbk)
ISBN: 9781041024255 (pbk)
ISBN: 9781003619178 (ebk)

DOI: 10.1201/9781003619178

Typeset in Times
by Newgen Publishing UK

Contents

Author's note vii
Preface ix
Introduction xi
About the Authors xv

Chapter 1 1892–1947: Early Materials Science 1

Chapter 2 1946: Flash of Brilliance 5

Chapter 3 1951: Purple Rising 7

Chapter 4 1964: Identification and Limitation 16

Chapter 5 1965: Purple Setting 20

Chapter 6 1970: First Detailed Failure Analysis and Metallurgy 28

Chapter 7 1971: IMC Formation Governed by Fickian Diffusion 32

Chapter 8 1976: Bond Shear Era Begins 44

Chapter 9 1982: Thermosonic Supplants Thermocompression – Active Intermetallic Compound Suppression 52

Chapter 10 1986: Take a Bromide? 58

Chapter 11 2000: Electrostatic Discharge Under Pad and Probe Marks 81

Chapter 12 2006: Impurities and the Reliability of Gold Wire 91

Chapter 13 Assessment of Au-Al Bond Interface Integrity 115

Chapter 14 Summary 124

Chapter 15 Glossary 125

Index 129

Author's note

The introduction of computer chips into consumer products necessitated plastic packaging that led to the introduction of gold wire bonding. The development of gold wire bonds to aluminum pads on electronic chips revealed many challenges. Brilliant minds across the world jumped on the opportunity to address the manufacturing and reliability challenges. Scientists and engineers evaluated the processes and materials and made advancements that enabled today's cell phones and other state-of-the-art technologies.

As a young man, I worked at Intel from 1979 to 1989 alongside many brilliant minds. It was an incredible learning experience, but no one taught me more than Rich Blish. His lessons extended well beyond technology.

The paper we (Ahmad, Blish, Corbett, King and Shirley) published in 1986 which detailed the degradation of Au-Al interconnects raised some questions that I continued to investigate, and in the process, I collected information on similar scientific phenomenon. I realized that I had collected information that is very useful for inquisitive people, as a result, I decided to put it in the form of a short book for the benefit of curious minds. Rich helped in giving it some structure and flow, and provided valuable advice. I am also grateful to Tom Dory and others who reviewed the manuscript and provided useful input towards its improvement.

Without the encouragement and support of my wife, Syeda Bushra Sultana, this compilation would not have been possible.

Syed Sajid Ahmad

Preface

Gold ball bonds to aluminum pads on integrated circuits have remained reliable for decades, with significant advancements in both reliability and physical understanding over time. Before the advent of automated wire bonders, operators were required to manually place each ball on a pad and position the silver plated leadframe for the stitch bond, an approach that naturally introduced opportunities for error and variation. Once the early wire bonders were "taught" the interconnect and leadframe patterns, then all bonds could be quickly and accurately placed. Early wire bonders used "thermocompression" bonding, in which a spherical gold ball; created by a hydrogen torch cutting the wire free from the previous stitch bond, was pressed onto a heated integrated circuit (IC) die. The stitch bond is inherently reliable as the gold-silver phase diagram shows complete miscibility and no intermetallic compounds. The Au-Al ball bond is completely different as its phase diagram shows several brittle, high resistivity (due to voids produced by the Kirkendall Effect) intermetallic compounds. Thermocompression bonds were driven strongly to end of life even during manufacture. "Thermosonic" bonds were (and still are) a substantial improvement over thermocompression, as the IC temperature could be reduced substantially, avoiding premature formation of lethal intermetallic compounds, which could die due to poor mechanical strength and high (notably, noisy) resistivity.

Other reasons for the longevity of Au-Al wire bonds are flexibility and cost, although there are many other interconnection schemes now available.

Richard C. Blish II

Introduction

Gold-aluminum interconnect has been an integral part of conventional chip packaging for decades. The gold-aluminum interface deteriorates during the operation of a device due to the formation of gold-aluminum intermetallic compounds. Interdiffusion during intermetallic formation causes vacancies which condense into voids, separating the interface, ultimately resulting in the catastrophic failure of the bond, and hence the device. With increasing device densities and overall package miniaturization, device heat dissipation densities are increasing, thus, adequate understanding of the phenomena to assure appropriate device life for the intended application is necessitated. Although the use of aluminum metal for on-chip interconnects in semiconductor devices is a widely used and mature process, it is valuable to understand the processes and implications from first materials principles.

Historical work to understand gold-aluminum wire-bond degradation is summarized. Salient excerpts from the several citations are provided here to facilitate the interpretation, without requiring the reader to chase down the references.

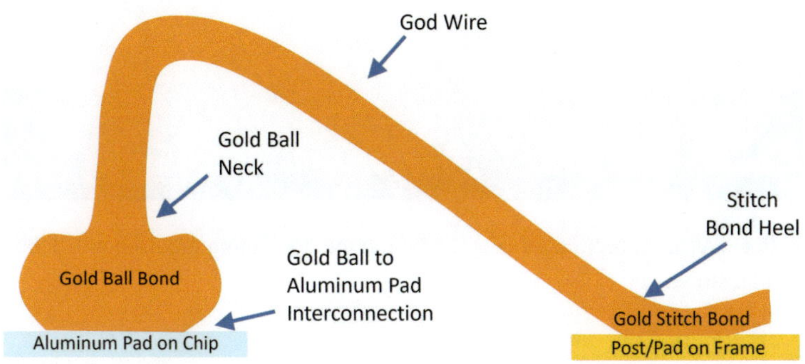

FIGURE 0.1 A conventional gold wire bond.

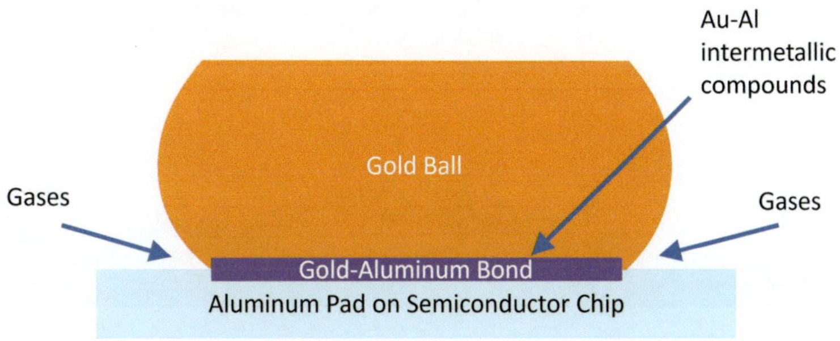

FIGURE 0.2 Gold-aluminum interface mechanics.

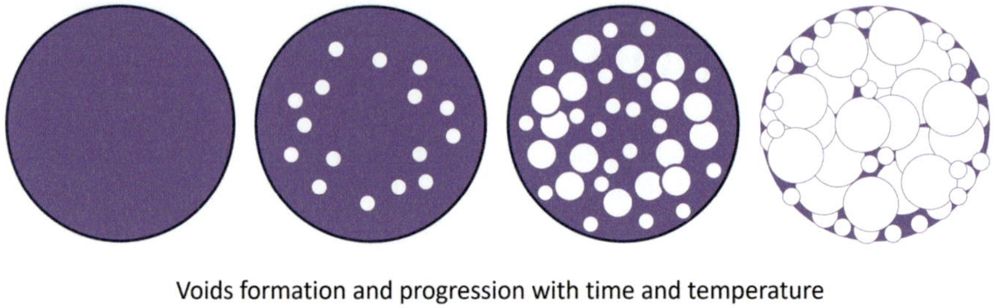

FIGURE 0.3 Voids formation and progression at the gold-aluminum interface.

FIGURE 0.4 A cross-section of a gold ball bond to aluminum As-bonded before any environmental stress.

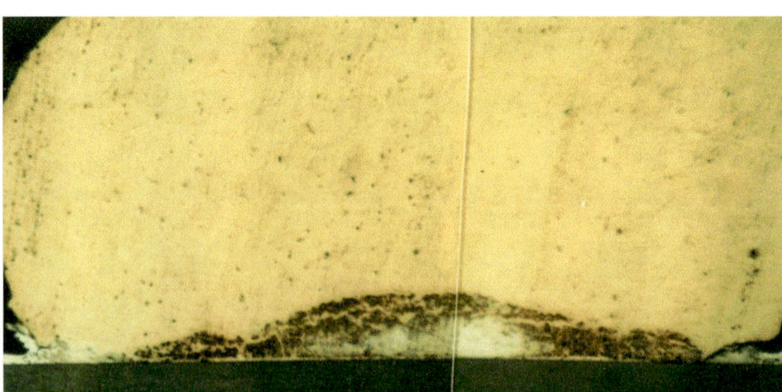

FIGURE 0.5 A cross-section of a gold ball bond to aluminum after aging under elevated temperature causing the formation and growth of intermetallic compounds at the gold-aluminum interface.

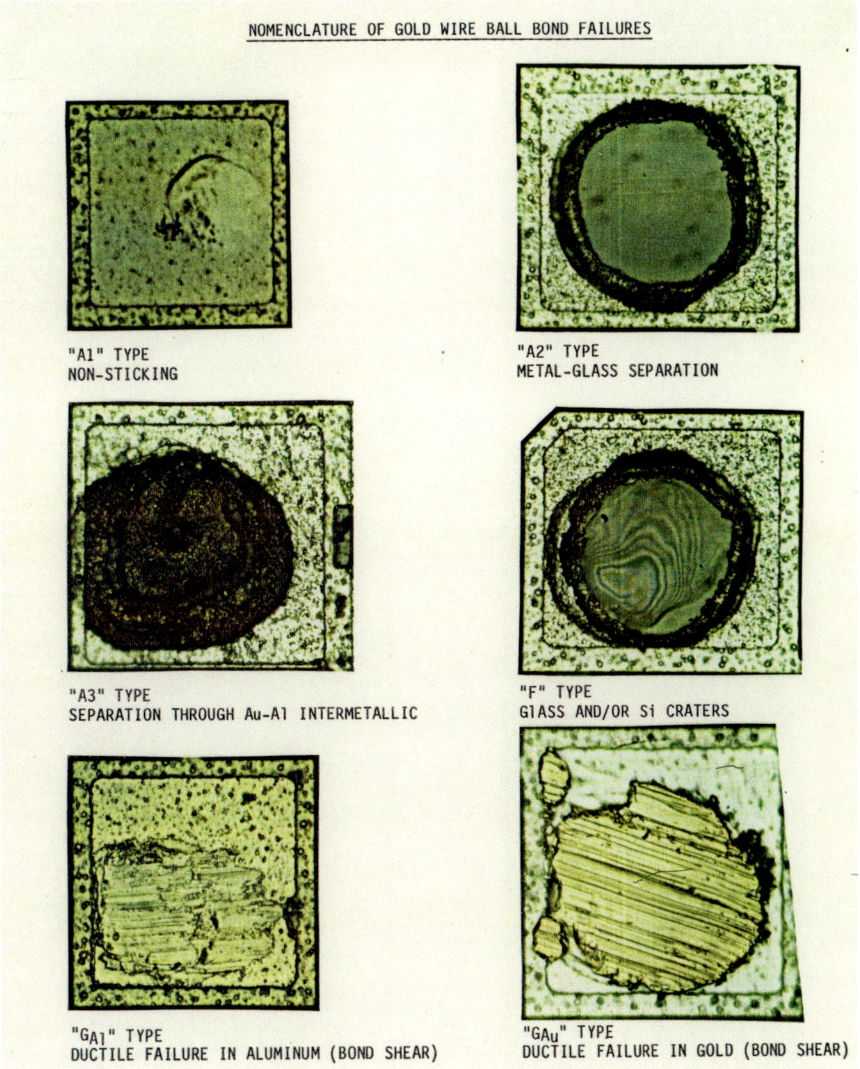

FIGURE 0.6 A nomenclature scheme for gold wire bond failures at the gold-aluminum interface shows various kinds of failures at gold wire pull or ball bond shear indicating quality and reliability status at the interface.

About the Authors

Syed Sajid Ahmad, Ph.D., contributed to quality and reliability enhancement of assembly processes at Intel (1979–1989), especially wire bonding.

Ahmad was involved in semiconductor packaging development at National Semiconductor (1990) and managed quality at GigaBit/TriQuint (1990–1991), a GaAs device company.

His major work at Micron Technology (1991–2003) involved materials enhancement, resulting in high reliability products and the development and implementation of advanced packaging of semiconductors.

At the Center for Nanoscale Science and Engineering (2003–2015) based at the North Dakota State University, his focus was on enhancing research and manufacturing capabilities at the center in the areas of thin film, thick film, advanced packaging (CSP) and surface mount technology (SMT).

At Crossfire Technologies (2018–2019), Ahmad contributed to the development and enhancement of assembly processes.

He has over 40 publications and presentations and holds 54 US patents.

Dr. Rich C. Blish II is a retired veteran, having worked in semiconductor reliability physics for the entirety of his 45-year career.

All of his degrees are from Caltech, Pasadena: BS in Physics; MS and Ph.D in Materials Science in 1967.

Rich spent two years with Bell Laboratories, Murray Hill, NJ, on the PicturePhone project; eleven years with Signetics (subsidiary of Phillips), San Jose, CA, doing analytical chemistry; fifteen years with Intel, Santa Clara, CA, in Package Reliability; eleven years with Advanced Micro Devices (AMD), Sunnyvale, CA, as an AMD Fellow in reliability modeling; and was a Spansion Fellow.

He has been a prolific author, having written about 60 patents and about 60 peer-reviewed scientific papers, winning awards for paper quality at the International Reliability Physics Symposium (IRPS) and the ECTC. The papers (and some book chapters) are on a variety of topics, for example, failure analysis, metallurgy, and the physics of radiation damage.

Dr. Blish is a past General Chair and Board Chair for IRPS. He chaired the Sematech RTAB in 2000 and 2007 and has been an Editor of the IEEE Transactions on Device and Materials Reliability.

1 1892–1947: Early Materials Science

This chapter covers the discovery of the gold-aluminum purple phase and its early characterization that defined its formation, structure and properties even before the manufacture of intricate electronic circuitry started.

A purple substance made up of gold and aluminum has been reported for more than a hundred years. In 1892, Sir William Chandler Roberts-Austen (1843–1902) reported in the *Proceedings of the Royal Society* [1],

> "The author has already described and exhibited to the Society a new alloy of gold and aluminum, $AuAl_2$, which is remarkable for its intense purple colour." …
>
> "In fact, when workmen who are accustomed to melt gold on a large scale, attempt to melt this purple substance, they find it difficult to believe that they are dealing with a gold alloy, as it is so infusible." …
>
> "The purple alloy presents the only case, known to the author, of an alloy, free from mercury, having a higher melting point than that of the least fusible of its constituents, and he considers that this fact affords strong evidence of its being a true compound of gold and aluminum." …

And, how truly he declared,

> The gold-aluminum series is of unusual interest, and well deserves careful attention.

This series continues to be of interest to date.
In 1893, Heycock and Neville found that [2],

> Aluminum and gold appear to form the compound $AuAl_2$ when dissolved together in molten tin.

Editorial Note: Surprised that copious Ag-Sn and Au-Sn intermetallic compounds did not occlude the conclusion. Some ternary intermetallic compounds may exist as well.

In 1894, Heycock and Neville published further results about the composition of the compound [3],

DOI: 10.1201/9781003619178-1

FIGURE 1.1 Sir William Chandler Roberts-Austen (1843–1902) who first reported purple Au/Al compound.

> "… the gold and aluminum form a stable insoluble compound of the formula $AuAl_2$ …"
>
> "The precipitate $AuAl_2$, is, no doubt, the purple alloy discovered and examined by Professor Roberts-Austen. He finds for his alloy the formula $Al_2Au_{0.98}$. Our values deduced from the freezing point curves are … $Al_2Au_{0.9}$ … $Al_2Au_{0.95}$ …"

In 1900, Heycock and Neville published the results of their work, which focused on determining the phase diagram for gold-aluminum interaction [4].

> "… the freezing point method is combined with a microscopic study of the alloys…" … "… a body crystallizes … which … is probably Au_5Al_2, but may be Au_8Al_3."

Editorial Note: Amazing that they determined phases correctly before 1900!
Rose documented in 1906 [5],

> Gold and [a]luminum, [t]hese alloys were investigated by Roberts-Austen, and more recently by Heycock and Neville. Several true compounds of the two metals were proved to exist, the most remarkable and stable being Au_2Al, a hard white alloy with a freezing point of only 622 °C, and $AuAl_2$, Roberts-Austen's beautiful purple alloy which melts at a higher temperature than the fusing point of pure gold. This compound is always formed when mixtures of gold and aluminum are fused, and allowed to cool, provided that not more than about 90 per cent. of gold is present. The purple alloy is seen in patches on a white ground containing the excess of aluminum or of white compounds of gold and aluminum.

In 1934, West and Peterson determined [6],

> $AuAl_2$ is formed by the reaction of the elements in the liquid phase with the evolution of considerable heat; it solidifies at the unusually high temperature 1060 °C (Au melts at 1063 °C) to a crystalline purple solid. ... With this structure the distance Au-Al is 2.58 or considerably less than the sum of the radii of the atoms 2.87. The volume per atom, 17.8 A^3, is greater than that found in Au, 16.9, or in Al metal, 16.5. … The purple compound $AuAl_2$ has most probably the CaF_2 structure with a = 6.00.

Editorial Note: By 1934, X-ray diffraction was available to determine crystal lattice dimensions.

In 1937, Zintl, Harder and Haucke studied the structure of $AuAl_2$ and determined that the blue colored composition of the compound contains a fluorite lattice. In view of these findings, Zintl et al. (1937) reported that [7],

> Alloy phases with a fluorite structure.

A subsequent study by Coffinberry in 1938, established that [8],

> solid solubility depends on the concentration of free electrons in the alloy, on the relative radii of solvent and solute atoms, on the forces of attraction between atoms, and on the relative stability of intermediate phases. The well-known Hume-Rothery rule states that the crystal structures and compositions of many intermediate phases depend on the electron-to-atom ratio.

They show their results in the in a phase diagram, combined with Heycock and Neville's liquidus and solidus curves. They took the diagram above 500 °C from Heycock and Neville's work.

> "… brilliant purple phase has … an extraordinarily high melting point, even higher than metallic gold. It is quite brittle. There … is no appreciable range of homogeneity." …

> "An unusually strong attraction between gold and aluminum atoms is indicated by the large negative deviation from Vegard's law, … by the evolution of heat on mixing, and by the stability of intermediate phases, especially $AuAl_2$. This suggests that these forces are not purely metallic but have a considerable ionic or covalent component. Brittleness and small range of homogeneity of intermediate phases bear this out." …

> "Our low value of 1.3 for the maximum electron concentration indicates the influence of other factors. A possibility is that the strong attraction of aluminum has increased the valence of some of the gold atoms above one. This is definitely indicated for the compound $AuAl_2$ … and is also indicated by generally low electron concentrations throughout the system."

> "The calcium fluoride structure of the eta phase, $AuAl_2$ resembles other compounds Mg_2Si, Mg_2Ge, etc., all with eight electrons to three atoms. This electron concentration is also derived from wave mechanics for this structure.

> This suggests that gold may have a valence of two in this case instead of the usual value, one. Such an increased valence of gold would not be unreasonable in a field of highly attractive aluminum atoms, which may tend to pull one of the inner electrons of gold into the outer shell." …
> "The very small solubility of gold in aluminum is probably due to the great stability of $AuAl_2$."

In 1939, Linde presented his thesis on the formulation of the resistivity of a binary alloy [9].

REFERENCES

[1] W. C. Roberts-Austen, "On the Melting Points of the Gold-Aluminium Series of Alloys," *Proceedings of the Royal Society of London,* vol. 50, pp. 367–368, 1892.

[2] C. T. Heycock and F. H. Neville, "The Freezing Points of Triple Alloys" (Chemical Society, December 21, 1893), *Nature,* vol. 65, no. January 25, p. 306, 1894.

[3] C. T. Heycock and F. H. Neville, "The Freezing Point of Triple Alloys," *Journal of Chemical Society, Transactions,* vol. 65, pp. 65–76, 1894.

[4] C. T. Heycock and F. H. Neville, "Gold-Aluminium Alloys," *Philosophical Transactions of the Royal Society of London. Series A, Containing Papers of a Mathematical or Physical Character,* vol. 194, pp. 201–232, 1900.

[5] K. Rose, *The Metallurgy of Gold*, Fifth Edition, Charles Griffin & Company, Limited: London, 1906, p. 14.

[6] C. D. West and A. W. Peterson, "The Crystal Structure of $AuAl_2$," *Zeitschrift fur Kristallographie, Kristallgeometrie, Kristallphysik, Kristallchemie,* vol. 88, pp. 93–94, 1934.

[7] E. Zintl, A. Harder and W. Haucke, "Legierungsphasen mit Flnoritstrnktnr," *Zeitschrift fur physikalische Chemie. Chemie der Elementarprozesse aufuau der Materie*, vol. 35, pp. 354–362, 1937.

[8] A. S. Coffinberry and R. Hultgren, "The Gold-aluminum System," *Transactions of the American Institute of Mining and Metallurgical Engineers,* vol. 128, pp. 249, 256–257, 1938.

[9] J. Linde, "*Thesis,* Stockholm: Cited in A. N. Gerritsen and J. O. Linde," *Physica*, vol. 18, no. 1952, p. 877, 1939.

2 1946: Flash of Brilliance

This chapter, while short, deserves its own place in this publication as it discusses the discovery that helped in understanding the gold-aluminum failure mechanism, even though the research was not directly related to the gold-aluminum interface study. The development of technologies surely relies upon scientific knowledge and its advancement.

In a meeting that took place in Atlantic City during November 1946 (published in 1947), Alice D. Smigelskas and Earnest O. Kirkendall (1914–2005) reported an observed interface shift and void formation in alloys or intermetallic compounds [1]. Editorial Note: This work was so controversial that it took Kirkendall a decade to get the discovery published, but in retrospect, asymmetric diffusion was the most significant metallurgy of the century for application to Au-Al wire bonding.

In 1997, Hideo Nakajima published an overview of Earnest Kirkendall's work regarding the Kirkendall Effect. [2].

> In 1947, Ernest Kirkendall reported the results of experiments on the interdiffusion between copper and zinc in brass and observed the movement of the interface between the different phases due to high-temperature interdiffusion, now called the Kirkendall Effect. This phenomenon supported the idea that atomic diffusion occurs through vacancy exchange. Since its discovery, the Kirkendall Effect has been found in various alloy systems, and studies on lattice defects and diffusion developed significantly. The Kirkendall Effect is important in connection with bonding between different materials and, in particular, raises the practical concern of controlling and suppressing the voids that are produced in the boundary region at a bonding interface. Today, the effect has been taken into account in various fields in materials science and technology such as structural materials welding, metals and ceramics powders sintering, thin films, and large-scale integration.

DOI: 10.1201/9781003619178-2

FIGURE 2.1 Earnest O. Kirkendall.

REFERENCES

[1] A. D. Smigelskas and E. O. Kirkendall, "Zinc Diffusion in Alpha Brass," *Transactions of the American Institute Mining and Metallurgical Engineers,* vol. 171, pp. 130–142, 1947.

[2] H. Nakajima, "The Discovery and Acceptance of the Kirkendall Effect: The Result of a Short Research Career," *JOM,* vol. 49, no. 6, pp. 15–19, 1997.

3 1951: Purple Rising

The beautiful purple is first mentioned in the analyses of the gold-aluminum interface, in relation to voids and change in resistance at the joint. Over time, change in resistance has become a useful tool that acts as an indicator of changes that occur at the interfacial level, and allows for the measurement of pore progression at the interface.

William Shockley, John Bardeen and Walter Brattain successfully demonstrated the transistor on 23 December 1947 at Bell Laboratories in Murray Hill, New Jersey. This opened the path for a new technology that needs interconnections between components, subsequently leading to the study of wire bonding and joining of different materials to make these interconnections possible, and understanding the interactions at material interfaces.

In 1952, Gerritsen and Linde noted change in the resistance behavior of a noble alloys [1],

> A number of alloys of a noble metal and a transition metal show anomalies in the resistance-temperature curve which differ essentially from the normal resistance-temperature relations which are found for pure metals.

In 1953, Seitz commented the following about the creation of voids in intermetallics [2],

> the work of the other investigators in the intervening period has indicated fairly conclusively that the porosity observed by Smigelskas and Kirkendall probably is a general phenomenon which must be regarded as an essential part of the diffusion process.

Seitz (1953) reviewed the explanations of the mechanism of void formation but admitted that they leave "many basic questions either unanswered or in little more than a qualitative state" and presented some of them.

In 1957, Anderson, Christensen and Andreatch described a "thermocompression bond" to a clean semiconductor surface below eutectic temperature. They claim to have observed no diffusion [3].

In 1959, Wittry reported the "use of electron probe X-ray microanalyzer" to confirm the presence of $AuAl_2$ in distended porous purple phase in attachments exhibiting poor mechanical strength when "leads, consisting of silver clad with platinum, were

DOI: 10.1201/9781003619178-3

attached by fusing with a gold-tin bead to the aluminum-silicon eutectic region on the transistors." [4]. He details the following,

> The electron probe microanalyzer has proved useful in the solution of a problem of this nature which arose in device development work, and which was called to our attention by L. Bernstein of Hughes Semiconductors. The problem was to obtain good mechanical strength in the attachment of leads to a silicon transistor. The leads, consisting of silver clad with platinum, were attached by fusing with a gold-tin bead to the aluminum-silicon eutectic region on the transistors. Many of the attachments exhibited poor mechanical strength which resulted in a high percentage of reject transistors. When the reject transistors were sectioned, a purple phase was observed next to the aluminum-silicon eutectic. In extreme cases, this phase often appeared *distended and somewhat porous* ... It was suspected that the purple phase was $AuAl_2$and to confirm this, a microanalysis was made in which the specimen was continuously scanned and the counting rate for the gold L_α doublet recorded. The specimen was then scanned in the reverse direction while the counting rate for the tin L_α line was recorded. ... It will be noted that where the separation between lead attachment and transistor occurred, the intensity of the gold line is equal to that from the standard of $AuAl_2$. At other points within the fused bead the intensity of the gold line corresponds to that from Au_2Al. The tin concentration of the phase richest in gold is very low while that in the purple phase may be as high as 5 weight-%.

Editorial Note: An important observation is that purple phase looked porous.

> The microanalysis of [successful transistor lead attachment] ... shows that no $AuAl_2$ was formed, although a phase having approximately the same gold content as Au_2Al is present in large quantities near the bead-transistor interface and is distributed throughout the bead in the form of a network. ... the gold-rich phase has almost no tin.

In 1959, as mentioned in Philofsky's paper, "Purple Plague Revisited," Bernstein reported seeing only $AuAl_2$ and Au_2Al in "synthesized" wire bonds [5].

In 1960, Domenicali presented a modeling of reaction mechanisms [6].

In 1960, Belser published a study of reaction mechanisms observing resistance change in Al-Au couples with temperature exposure and observed the presence of $AuAl_2$ in X-ray diffraction studies [7].

> Regardless of the method of coating, i.e., evaporation, sputtering, or electroplating, films were found to interdiffuse or alloy at or below the temperature of recrystallization of the metal of higher melting point.

Belser found that the resistance change did not follow the expected rate of increase over time and temperature.

In 1961, Bernstein [8] and [9] reported,

> "Wire or preforms of pure gold, pure gold with added dopants, Au-Si, Au-Ge, and Au-Sn were alloyed onto Al-Si eutectic at temperatures in excess of

400 °C. In every case, metallographic sections disclosed the presence of the "purple plague," as well as a second "gold-ish" phase, along with the original Au and Al-Si phases. X-ray diffraction and spectrographic analysis identified the new phases as $AuAl_2$ and Au_2Al, respectively."

"It is seen that elevated temperatures are required to produce the gold-aluminum alloys from the elements. ... A liquid phase which would permit mixing of Au and Al for compound formation during electrode attachments between Au and Al-Si eutectic would not be expected at any temperature below the eutectic melting point at 577 °C. It was, therefore, interesting to find the "purple plague" was actually produced at temperatures as low as 425 °C. No bond was successfully formed below 425 °C. This was true so long as Au, Al and Si only were present. ... At temperatures of up to 500 °C for one minute there was no appreciable increase in the amount of $AuAl_2$ formation."

"The interface is porous and discontinuous with resultant cracking of the silicon."

"The mechanism for these reactions is believed to be as follows: In the three component Au-Al-Si system there can exist a liquid phase above 370 °C (Au-Si eutectic temperature). This liquid containing gold and silicon can dissolve aluminum, thereby raising its melting point and freezing, or form a still lower melting ternary eutectic. In either case, if we consider non-equilibrium conditions, both gold-aluminum compounds can form. Evidence of the solubility of Si in Au-Al has been found by X-ray diffraction and spectrographic analysis. In addition Si dispersed both in the purple and the gold-ish phases in the course of electrode attachment has been detected in a metallographic section. This silicon in contact with more gold at temperatures above 370 °C produces more liquid, thereby continuing the cycle until equilibrium is reached or the reaction arrested by lowering the temperature."

"Generally speaking, the presence of the purple $AuAl_2$ as a thin film impairs neither the mechanical strength nor the electrical properties of semiconductor devices when operated at room temperatures and low current levels. There is reason to believe, however, that some devices operating at high temperatures and/or current levels may be adversely affected electrically by the presence of even a thin film. The sensitivity of the metallurgical reactions to time, the co-existence of the two gold-aluminum compounds, and the presence of the five phases in this three component system all point out the non-equilibrium conditions under which this attachment is normally made."

"Because of the great differences between the expansion properties of Al-Si and Si, junctions formed by alloy fusion sometimes produce cracking of the silicon. By carefully controlling the junction geometries, thickness ratios of materials and cooling cycle, the danger of cracking, or bowing for that matter, can be completely eliminated."

In 1961, Kidson discussed certain aspects pertaining to the growth of diffusion layers in binary systems [10],

> The application of Fick's first law to polyphase diffusion in binary systems leads to a simple physical interpretation of various possible types of behavior in the growth of diffusion layers. … there is no justification for assuming that each phase in the equilibrium diagram of a system will grow to an observable width in a diffusion couple.

Editorial Note: True, and diffusion may not even be Fickian (i.e., phase thickness may not be proportional to the square root of time. Some researchers have also indicated that their data did not fit Fick's model).

In 1961, Miller and Mackintosh presented an example of metallurgical reactions affecting contact resistance [11]. They show thermocompression bonds between the alloyed aluminum contact on a diffused silicon contact and the gold lead wire surrounded by a discoloration due to the presence of a compound formed when gold-aluminum joint is heated between 200–300 °C for extended periods of time contributing to the deterioration of the contact.

Schmidt (1962) on the other hand, postulates that silicon acts as a catalyst in causing the formation of Au_2Al [12].

> This paper reports the aging behavior of semiconductor contacts which are made by thermocompression bonding of a gold wire to an aluminum stripe. These contacts degrade quite rapidly during aging of the semiconductor devices at temperatures above 150 °C. The resistance between the lead wire and semiconductor increases with aging time and quite often a complete opening of the contact is experienced. These effects are caused by the formation of a compound with a black surface which spreads out from the gold wire bond over the aluminum stripe. The color of this compound below the surface varies from a faint yellow to a dirty white or silver. The lead wire shows a discoloration in the bond area and a black material in the grain boundaries of the flattened wire. Quite often a loss of gold is evidenced by holes in the flattened wire. The compound forms in all ambients from room air to vacuum of the order of 10^{-5} torr. Evidence is presented to show that
>
> 1) The degradation is caused by the formation by a solid-state reaction of a compound consisting mainly of aluminum and gold and the reaction rate is not controlled by bulk diffusion
> 2) The compound can form only, (a) if gold and aluminum are in metallic contact, or (b) if both have access to a semiconductor surface and are in diffusion range of each other [i.e., silicon acts as a catalyst].
> 3) The gold lead wire is weakened either by gold diffusing out of the bond area or by grain boundary diffusion of the black compound between the gold crystals.
> 4) Whenever gold and aluminum are in metallic contact, the purple compound $AuAl_2$ will form by bulk diffusion. This compound is stable, is a true metal and therefore a reasonably good conductor and, while brittle, is reasonably strong. Therefore, the formation of this material does not explain the observed effects.

Editorial Note: The expression "reasonably good conductor" is too optimistic … Typically a fully dense intermetallic compound will have a conductivity roughly an order of magnitude larger than that of its pure elements, whereas an intermetallic compound made porous by Kirkendall voiding can exhibit a resistivity up to four orders of magnitude higher than that of the pure elements, as shown in the work of Blish et al. in 1983.

In 1962, Dodson reported his observations on gold-aluminum bonds made on germanium and noted their degradation at elevated temperatures [13].

> This paper describes the thermal aging behavior of metal-aluminum bonds made to germanium devices. Specially prepared test structures of gold, copper, silver and platinum wire bonded to aluminum stripes over a germanium substrate have been used. The techniques of accelerated aging have been used to characterize the aging behavior of these bonds in the 200–350 °C temperature range.
>
> The aging characteristics are revealed in two ways. 1) The ohmic resistance of the bonds increases with time (0.1-5 ohms). 2) The bonds are physically weakened, becoming intermittent and then completely open circuits.
>
> Although the aging behavior for each of these metal-aluminum bonds is similar, the time rate of change of the resistance and therefore the failure mechanism may be different. However, for all cases, unsealed units show a longer time for a 2-ohm resistance change by a factor of 10-1000 than do units sealed in room air.
>
> When the contacts become an open circuit, the wire generally lifts out a small chunk of material from the germanium substrate leaving a hole 1-2 microns deep.

Dodson (1962) noted that gold-aluminum bonds made on germanium exhibited an anomalously resistant degradation rate at elevated temperatures.

Weaver and Brown, reported in 1962 [14],

> Diffusion coefficients and activation energies for diffusion in successively evaporated films of gold and aluminum have been determined by observing the changes in reflectivity at the gold and aluminum surfaces on annealing. The changes were characteristic of a sharply defined reaction boundary rather than of a gradual change in concentration. The variations with film thickness showed that this was a phase boundary caused by the formation of the intermetallic compound Au_2Al. The presence of this compound was confirmed by electron diffraction. Results indicated a vacancy diffusion mechanism rather than grain boundary diffusion.

Editorial Note: This marks the first explicit proof of void mechanism dominating grain boundaries.

> "… very thin gold films … were aged over the temperature range 20–95 °C, and the activation energy of diffusion was found to be 23.3 kcal/mole, in excellent agreement with the values for thicker films (22.6 kcal/mole, [0.98 eV/atom])."

"… very thick gold substrate films … were aged over a temperature range from 70 °C to 152 °C. … the activation energy was calculated to be 23.5 kcal/mole [1.02 eV/atom]."

"The activation energy at the gold surface was found to be 22.6 kcal/mole whilst that for diffusion into the aluminum was found to be 23.5 kcal/mole."

"… Reflectivity measurements at both the gold and the aluminum surfaces seem to indicate the motion of a particular phase boundary, and in view of the critical thickness ratios which have been found, it would appear that this corresponds to formation of a layer of Au_2Al."

"… the penetration of the diffusion layer into the gold is 2.3 times greater than its penetration into the aluminum."

Editorial Note: Diffusion speed differential is characteristic of Kirkendall voiding

In 1962, Phillips, Anderson and Erickson reported studies on fundamental failure mechanisms [15].

"In silicon devices these stripes are usually made of aluminum deposited to a thickness of approximately 20 to 40 microinches. A common crystal and terminal "post" configuration on a TO-18 header then results in internal leads about 30 to 40 mils long, and the dimensions of the stripes dictate a lead diameter on the order of 0.5 to 0.75 mils. Added to these rather stringent mechanical considerations are the problems associated with thermocompression bonding, the technique used in attaching the leads to both stripes and terminal posts. It is interesting, therefore, that virtually all suppliers faced with these problems settled on gold as the lead material which would satisfy all requirements, and indeed a reasonable review of its chemical and physical properties would tend to support their choice."

"Early in our evaluation of silicon mesa and planar transistors, [we experienced] catastrophic failures due to internal opens, [which] began to appear, and, nearly always, separated bonds or "broken" lead wires right at the bond were involved. At first it was felt that the difficult bonding process[es] might be out of control. [In this scenario, the process] could [have potentially produced] bonds that did not [combine or] stick properly, [due to insufficient] pressure [applied] on the bonding tool or [an incorrect substrate temperature] at the time of bonding. [Conversely,] it could also result in leads that were all but sheared off [due to excessive] pressure and/or improper [poor] bonding tool design."

"Invariably, however, our observations of these failed units showed the presence of dark areas on the aluminum stripe in the region surrounding the bonds. The significance of this condition was not understood or explained until metallurgists identified it as evidence of an intermetallic reaction between gold and aluminum. This reaction is chemical in nature and is promoted by the relatively high temperatures used in the bonding and stabilization processes for silicon devices. Of the various intermetallic compounds formed, $AuAl_2$ is the worst. [A few of] its undesirable properties [include:] poor electrical conductivity, brittleness and a coefficient of thermal expansion different from either gold or aluminum. The condition became widely recognized, and even acquired the disagreeable appellation of "purple plague" now commonly used in the industry."

In 1963, Jan and Pearson reported [16],

> $AuAl_2$ which has a brilliant mauve colour, appears to be a rather good metal as far as its electrical properties are concerned, since its room temperature resistivity is only about five times as large as that of copper and its residual resistance ratio is not much more than 1/200.

Editorial Note: The residual resistance ratio is the quotient of resistivity at ambient temperature to that at cryogenic temperature, serving as an index of defect density. While AuAl would be a "good metal" if it were free of Kirkendall voids, the voids will always be present in $AuAl_2$, degrading resistivity by ~3 orders of magnitude.

> When we compare this with the electrical behaviour of metallic superstructures of the Cu_3Au or CuAu type (cf. e.g. Hirabayashi and Muto 1961, Anquetil 1962) that exist over a range of composition, it seems that metallic 'compounds' such as AuAl, must attain a high degree of perfection in the location of the atoms on the crystal lattice. They are probably only homogeneous at a unique composition.

Longo and Selikson (1963) reported the following in their work on the identification and elimination of a failure mechanism in semiconductor devices [17],

> "The use of high temperature materials for silicon transistor fabrication is desirable in order to capitalize on the 200 °C junction operating temperatures which are inherently possible. Accelerated life testing in reliability studies are therefore usually conducted at 300 °C. for hundreds and even thousands of hours. Process stabilization bakes are also usually performed at 300 °C."
>
> "… The elimination of either the gold wire or the aluminum metallization also presents some rather tricky problems. Gold metallization, for example, does not lead to a satisfactory solution. Silicon has a very strong affinity for oxygen and, therefore, is always covered with a thin oxide layer. The gold metallization frequently does not penetrate the thin oxide to alloy into the silicon. Aluminum, however, has an even stronger affinity for oxygen than has silicon and can therefore penetrate the thin oxide layer and alloy into the silicon readily.
>
> "… Silver when used as a lead wire has been observed to exhibit a peculiar dendritic growth under certain ambient conditions. Such dendrites could, of course, cause shorting between emitter and base due to the very close spacings involved. Also, a degradation of the strength of contacts bonded between silver wires and aluminum metallization has been observed."

Editorial Note: Moisture and bias polarity are critical for silver dendrite growth.

In another article published by Longo and Selikson (1963) reported [18],

> The purpose of this article is to identify the failure mode as the formation of a ternary aluminum-gold semiconductor black compound or complex. The failure mode is eliminated by the use of aluminum wire bonding to the semiconductor.

That is, the black compound is caused by the presence of gold as it does not appear in aluminum bonding.

The same year, in 1963, a series of reports from RCA on *Production Engineering Measure on 2N1708 Silicon Planar Epitaxial Transistor* elicited that the "use of thin aluminum contacts ... inhibited formation of "purple plague." ... The use of gold alloy wire did not materially improve the bond strength." [19, 20].

The following year, in 1964, Adams and Workman detailed open failures due to purple plague and mentioned a method to identify the open failures using a curve tracer. They wrote [21],

> The technique of looking for open bonds consists of forward-biasing the substrate of the network bar with respect to the other elements of the bar, using a curve tracer. The resultant forward diode characteristic is the clue to an open bond. If a bond to the substrate or other element involved is open, the trace is a horizontal line. If the trace is that of a normal forward diode or diodes, the bonds involved are intact... The bonds had opened as a result of an inherent chemical reaction between aluminum and gold. The resulting compound or compounds—commonly referred to as "purple plague"—are brittle. The bonds are normally under tension as a result of wire tautness; therefore, this force tends to break the brittle material which, in turn, leads to an opening. In addition to wire tension, there appears to be a volume increase as the reaction progresses. This increase is nonuniform across the bonded interface and pushes the non-plaque area apart.

They reported acceleration factors of 50, 18.7, 6.8, 2.5, 1 and 0.37 at 300, 200, 125, 85, 55 and 25 °C respectively that yields an activation energy value of just 0.18 eV.

REFERENCES

[1] A. N. Gerritsen and J. O. Linde, "The Electrical Resistances of Alolys of a Noble Metal and a Transition Metal," *Physica,* vol. 18, no. 11, pp. 877–890, 1952.

[2] F. Seitz, "On the Porosity Observed in the Kirkendall Effect," *Acta Metallurgica,* vol. 1, no. 3, pp. 355–369, May 1953.

[3] O. L. Anderson, H. Christensen and P. Andreatch, "Technique for Connecting Electrical Leads to Semiconductors," *Journal of Applied Physics,* vol. 28, no. 8 (August), p. 923, 1957.

[4] D. B. Wittry, J. M. Axelrod and J. O. McCaldin, "Use of the Electron Probe X-Ray Microanalyzer in the Study of Semiconductor Alloys," in *Semiconductor Conference, AIME*, Boston, August 31–September 2, pp. 89–100, 1959.

[5] L. Bernstein, *Semiconductor Products,* p. 34, 1959 [Philofsky, Purple Plague Revisited, 1970].

[6] C. A. Domenicali, "Thermoelectricity and Resistivity in Metal Alloys at Low Temperatures," *Physical Review,* vol. 117, no. 4 (February 15, 1960), pp. 984–992, 1960.

[7] R. B. Belser, "Alloying Behavior of Thin Bimetal Films, Simultaneously or Successively Deposited," *Journal of Applied Physics,* vol. 31, no. 3 (March), p. 562, 1960.
[8] L. Bernstein, "Gold Alloying to Germanium, Silicon and Aluminum-Silicon Eutectic Surfaces Part 1," *Semiconductor Products,* vol. 4, no. 7 (July), pp. 29–32, 1961.
[9] L. Bernstein, "Gold Alloying to Germanium, Silicon and Aluminum-Silicon Eutectic Surfaces Part 2," *Semiconductor Products,* vol. 4, no. 8 (August), pp. 35–39, 1961.
[10] G. V. Kidson, "Some Aspects of the Growth of Diffusion Layers in Binary Systems," *Journal of Nuclear Materials,* vol. 3, no. 1, pp. 21–29, 1961.
[11] L. E. Miller and I. M. Mackintosh, "Semiconductor Reliability," in *Conference on Reliability of Semiconductor Devices*, Edited by John E. Schwop, Harold J, Sullivan, Engineering Publishers, 1961, pp. 107–107.
[12] R. Schmidt, "Mechanism of Lead Failures in Thermocompression Bonds," *IRE Transactions on Electron Devices,* vol. ED-9, no. November, p. 506, 1962.
[13] G. A. Dodson, "Aging Problems of Thermocompression Bonds Made to Semiconductor Devices," *IRE Transactions on Electron Devices,* vols ED-9, no. November, p. 506, 1962.
[14] C. Weaver and L. C. Brown, "Diffusion in Evaporated Films of Gold-Aluminium," *Philosophical Magazine,* vol. 7, no. 73, pp. 1–16, 1962.
[15] R. G Phillips, G. Anderson and R. Erickson, "Fundamental Failure Mechanism Studies," in *First Annual Symposium on the Physics of Failure in Electronics*, Chicago IL, 26–27 September 1962, pp. 73–90.
[16] J. P. Jan and W. B. Pearson, "Electrical Properties of $AuAl_2$, $AuGa_2$ and $Auln_2$," *The Philosophical Magazine,* vol 8, no. 86, pp. 279–284, 1963.
[17] T. A. Longo and B. Selikson, "Identification and Elimination of a Failure Mechanism in Semiconductor Devices," *Physics of Failure in Electronics*, vol. 2, pp. 424–435, 1963.
[18] T. A. Longo and B. Selikson, "Aluminum Wire Bonding of Silicon Transistors," *Semiconductor Products,* vol. 6, no. November, pp. 27–31, 1963.
[19] L. Possemato and W. Totten, *Production Engineering Measure On 2N1708 Silicon Planar Epitaxial Transistor*, RCA, Somerville NJ, 1963.
[20] W. Totten and L. Possemato, *Production Engineering Measure on 2N1708 Silicon Planar Epitaxial Transistor*, RCA, Somerville NJ, Period Covered 1 February 1963 through 30 April 1963.
[21] J. Adams and W. Workman, "Semiconductor Network Reliability Assessment," *Proceedings of the IEEE*, vol. 52, pp. 1624–1635, December 1964.

4 1964: Identification and Limitation

We continue to see how researchers struggled with assigning a mechanism for their observations and trying to make sense of their discoveries as they developed processes that aimed to increase the yield and reliability of the products they were developing.

The same year, in 1964, Longo and Selikson showed a substantial decrease in bond strength due to the phenomenon [1].

> "... the rate of formation of purple plague is catalyzed by the presence and interaction of silicon."

> "... Aluminum is a popular metallization material since it makes excellent ohmic contact to *p*-type and degenerate *n*-type regions in silicon, and because the evaporation, alloying and recrystallization of aluminum in silicon is familiar and relatively easy to control. Also, the high free energy of formation of aluminum oxide compared to that of silicon dioxide indicates that aluminum would reduce any residual surface oxide on the silicon, thereby making it available for uniform alloying. These properties make aluminum ideally suited as a silicon contact material."

> "To study the relative mechanical strength of the aluminum wire bonded units relative to the gold wire bonded units, as well as the relative rate of formation of purple plague in each unit, several hundred NPN silicon planar epitaxial transistors were chosen from one dice production and header assembly lot and half were bonded with gold wires and the other half with aluminum wires... Each group was divided into 4 subgroups and an aluminum and a gold wire group were each heated at 300 °C for 18 hours, 96 hours, and 425 hours, respectively. The aluminum wire bonded units showed a remarkable difference from the gold wire bonded units as to color, rate of development of color, and bond strength. On overnight baking, all gold wire units developed extensive black plague at and around the bonds on the semiconductor, while the aluminum wire bonded units only developed purple plague at the posts after hundreds to thousands of hours baking."

Through the progression of time, over the years, as experiments with Au-Al bonds are conducted, we will see that 300 °C is, in fact, quite a high temperature for these bonds to bear. It's no wonder the bond strength falls rapidly in the graph in

DOI: 10.1201/9781003619178-4

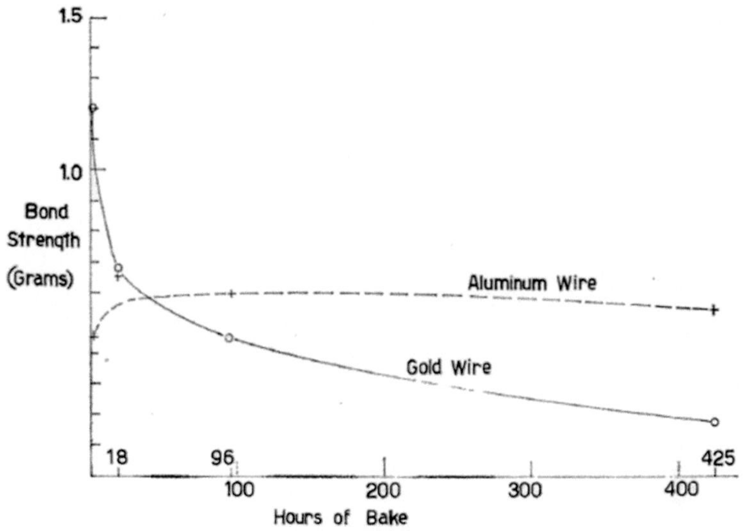

FIGURE 4.1 Bond strength versus. hours of baking for gold and aluminum wire bonded transistors.

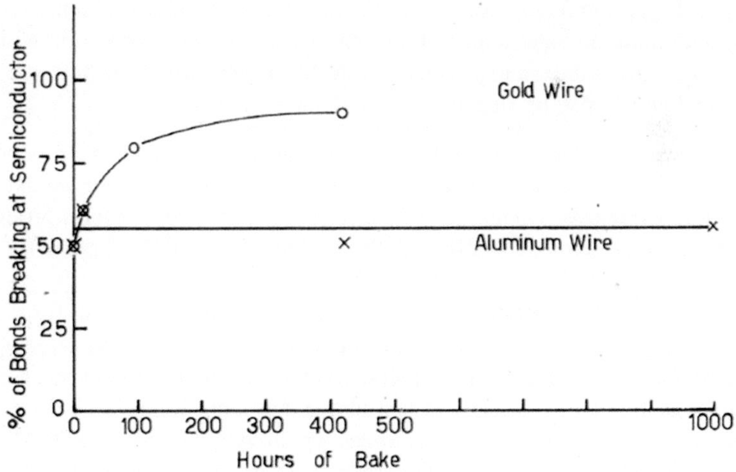

FIGURE 4.2 Percentage of bonds breaking at semiconductor versus hours of baking for gold and aluminum wire bond transistors.

Figure 4.1. Longo and Selikson's study also shows that bond failures are high in the graph in Figure 4.2, which was included in their report.

Longo et al. also mentioned purple and blackish (with Si) phases [2].

In 1964, Rush encountered purple plague in his solar flat thermoelectric generator research. [3].

> An electrodeposited layer of gold was applied to the aluminum foil so that the thermoelement assemblies could be attached, by soldering, to the foil. ... when [the] panel ... was exposed to thermal cycling ... [s]evere corrosion of the collector plate was observed at the end of the test. This was diagnosed as the formation of a gold/aluminum intermetallic compound ($AlAu_2$). [The] problem ... was resolved by substituting nickel for the gold electrodeposit.

In 1965, Bob G. Bender, and Leonard Bernstein obtained a patent on a method to alleviate the purple plague problem. The patent was originally filed in (Bender & Bernstein, 1959) [4]. They claimed,

> "An aluminum-silicon eutectic surface is produced in semiconductor device manufacture by alloying aluminum on to a silicon semiconductor crystal to produce a PN junction. Satisfactory junctions are easily produced in this manner, but the attachment of leads or electrodes to the predominantly aluminum surface phase is difficult... According to the present invention, a low melting alloy of gold and tin is used as a bonding alloy to bond leads to such predominantly aluminum surfaces without requiring the use of fluxes, or agitation. An alloy of gold and tin is used, [this is because it has a] melting point [which is] sufficiently low [enough] that the formation of the deleterious purple phase may be avoided. An eutectic mixture of about 80 % gold and 20 % tin by weight has a melting point [of] 280 °C and makes a very satisfactory bonding alloy. This predominantly gold bonding alloy will form a bond to a predominantly aluminum phase at temperatures above its 280 °C melting point without requiring a special flux, and produces an etch-resistant, strong, shallow penetration bond which may be cleaned without deleterious effects."

> "It has been found that when gold and aluminum are present in a bonding operation, a purple, brittle $AuAl_2$ phase may be formed. [When exposed to temperatures] above critical, this phase forms rapidly, and [notably] produces an unsatisfactory bond. Usually a small, discontinuous purple phase is formed at temperatures as low as 25 °C below the critical temperature, [however,] unless the critical temperature is exceeded, no such rapid phase transformation takes place. Aluminum and gold may be held just below the critical, rapid phase transformation temperature for several minutes without forming substantial volumes of the purple phase, but just above that temperature, a few seconds to minutes are sufficient to form volumes which would require one to several hours to form below the critical temperature. By holding the bonding temperature below the $AuAl_2$ rapid phase transformation temperature, deleterious formation of $AuAl_2$ is avoided."

Editorial Note: This is a plausible explanation regarding purple phase formation but is neither correct nor relevant. Regardless, this remains a valuable patent, and is still in use decades later as Be slows Au grain-growth above the ball, such that failures do not occur just above the ball. See James (1980) [5].

REFERENCES

[1] T. A. Longo and B. Selikson, "Identification and Elimination of a Failure Mechanism in Semiconductor Devices," *Second Annual Symposium on the Physics of Failure in Electronics,* 25–26 September 1963, Chicago IL, pp. 424–435, 1963.

[2] T. A. Longo and B. Selikson, "A Study of Purple Plague and Its Role in Integrated Circuits," *Proceedings of the IEEE,* vol. 52, no. December, pp. 1638–1641, 1964.

[3] R. E. Rush, *Solar Flat Plate Thermoelectric Generator Research*, General Instrument Corporation, Newark NJ, September 14, 1964.

[4] B. G. Bender and L. Bernstein, "Gold-Aluminum Bond Alloy Attachment," USA Patent 3202489, 24 August 1965.

[5] H. K. James, "Resolution of the Gold Wire Grain Growth, Failure Mechanism in Plastic Encapsulated Microelectronic Devices," *IEEE Transactions on Components, Hybrids, and Manufacturing Technology,* vol CHMT-3, no. 3, pp. 370–374, 1980.

5 1965: Purple Setting

In the continuous struggle to characterize mechanisms for gold-aluminum bond failures, bond resistance and gold wire composition also come into consideration.

In 1965, Cunningham suggests that the predominant intermetallic under gold ball bonds is Au_4Al causing substantial decrease in bond strength [1].

In 1965, while discussing ternary compound with Si, Ruggiero reported [2],

> Evidence is given to substantiate the hypothesis that purple plague ($AuAl_2$) has poor adhesion to SiO_2 and thereby induces premature failure of semiconductor devices whose gold wire leads are bonded to aluminum over SiO_2. Evidence is also given that, after aging, better adhesion of gold wire bonds to metalized microcircuits is achieved when the aluminum metallization is alloyed into silicon. Thus it appears that lack of adhesion, rather than brittleness of the purple plague, causes failure of gold wire bonded to aluminum metallization on microcircuits.

Editorial Note: Once full annealing, accounting for time at temperature during bonding, operational lifetime, or under accelerated conditions, completely consumes the Al film, adhesion is lost, as gold does not adhere to silicon.

> $AuAl_2$ was found to form in a similar manner on surfaces containing Si or SiO_2. The indication is that aluminum reacts with the SiO_2 to give silicon, and a ternary Si-Al-Au system forms which lacks adhesion to the surface. On surfaces lacking silicon, the gold-aluminum compound does not readily part from its substrate.

In 1965, Colteryahn and Shaffer reported [3],

> It is postulated that the formation and transformation of intermetallic compounds will be accompanied by formation of strained interfaces due to mismatch of atomic lattices and also changes in volume inducing stresses into the TC bond. The strained interfaces would be susceptible to fracture by stresses induced from aforementioned volume changes. It is known that the properties of intermetallic compounds generally are very different; they are hard and brittle.

DOI: 10.1201/9781003619178-5

Vaccaro and Gorton's (1965) Reliability Physics Notebook sums up knowledge of purple plague at the time as [4],

> Purple plague: [t]his descriptive term refers to a contact problem that develops between gold wire leads and aluminum metallization. It is now established … that this involves the growth of a brown or dark alloy of gold, aluminum, and silicon, referred to as the "black death," which is quite brittle, and hence mechanically weak, and which increases the electrical resistance of the leads. The term "purple plague" arose because the intermetallic compound $AuAl_2$ is purple. In apparently successful attempts to solve the problem of the "black death" alloy by using aluminum wires, it was observed that purple $AuAl_2$ developed where the aluminum wires were bonded to the gold-plated Kovar terminals, but nevertheless without any objectionable effects. The bond remained mechanically strong, and no significant increase in electrical contact resistance resulted.
>
> The growth rate, both of "purple plague" and "black death," is only appreciable at temperatures above 100 °C.

In 1966, Browning et al, in their study of failure mechanisms in thermocompression bonds in integrated circuits, confirmed the presence of only Au_4Al, Au_5Al_2 and Au_2Al in gold ball bonds using the electron microprobe. Interestingly, they found little evidence of "purple plague," but did convey that Kirkendall voiding was the cause of diagnosed failures [5].

Editorial Note: It is worth noting that while Au_4Al was first reported in 1966, there was NO mention of $AuAl_2$. Moreover, it is also important to remark that a massive Au ball on a thin Al film would be expected to favor the formation of Au-rich phases, however, this was not the case.

> "Intermittent behavior, however, was not recognized until it was observed that some open bonds closed when a static charge accumulated on the case of the device. Following this tenuous lead further, intermittent behavior was observed to occur rather frequently, was established to be dependent upon voltage impressed during electrical testing, was influenced by test temperature, and showed some time dependency. Some bonds could be made to behave intermittently almost at will, thus virtually making them act as a mechanical on-off switch."

> "… thorough microscopic examination of open and intermittently open bonds showed that a crack, void, or discontinuity of some kind was visible at the outer edges. There was some microscopic evidence that a discontinuity was entering the intermetallic band near mid-thickness rather than along the interface between the band and the unaffected gold mass."

> "… This work confirmed the presence of the previously predicted Au_4Al and Au_5Al_2 intermetallic compounds in the band and also the compound Au_2Al. It is interesting to note that very little, if any, evidence was found for the compounds AuAl, $AuAl_2$ (commonly known as "purple plague" among device manufacturers and users) or unreacted aluminum. This does not imply that AuAl, $AuAl_2$ or unreacted Al are not present but simply that, if present, they are beyond the resolution capability of the electron beam of the microprobe

> apparatus. Furthermore. no detectable amount of Si was found in any of the bonds scanned with the microprobe."

Blech and Sello (1966), in their work *Some New Aspects of Gold-Aluminum Bonds* reported [6],

> Gold-aluminum diffusion couples as well as gold thermocompression bonds onto aluminum films were carefully examined after aging. The samples show the presence of all intermetallic phases predicted by the phase diagram. It is shown that voids formed by the Kirkendall effect are responsible for bond failures in bulk gold-aluminum diffusion couples. While voids have been seen in Au ball bonds on thin Al films, these rarely contribute to mechanical failure. In the case of a very thin Al film on silicon oxide (0.05 μm), electrical failure (open) was observed due to the disappearance of Al, exposing the underlying oxide. None of the previously reported causes of bond failure accurately describe the aging process.
>
> "… twenty 1 mil gold ball bonds on transistors which averaged 5.9 gr initially showed 5.5 gr of pull after 1,000 h[ours] at 200 °C, all breaking in the gold wires showing a small loss in effective pull strength. This loss in pull strength is due to the softening of the gold wires on annealing… Electrical failures of AuAl bonds over oxide are serious when the Al films are too thin. The aluminum diffuses into the gold, leaving bare oxide.

Editorial Note: While the observation about thin Al is important, the notion of "softening of gold wire" or "softening" the gold is not convincing. As noted elsewhere, Al diffuses rapidly into Au, so a thin Al film can easily be isolated by a "moat."

Bert Brennen declared in his 1966 patent *Beryllium-Gold Alloy and Article Made Therefrom* [7],

> "… the addition of extremely small or minute amounts of beryllium to gold has the surprising, unexpected and highly desirable and beneficial effects of (1) increasing the tensile strength to a great extent, (2) substantially improving the ductility of gold in fine wire form and (3) rendering the gold stable. Hence, a beryllium-gold alloy according to this invention affords a relatively high purity gold which is both ductile and stable and which may be formed more readily and economically into fine wire than gold that does not contain beryllium in the amounts indicated above and identified more particularly hereinafter.

In 1966, Berger, Keen and Schnable published the title, *Reliability Phenomena in Aluminum Metallizations on Silicon Dioxide*, and wrote [8],

> "… lateral migration of aluminum toward the gold bond occurs thereby causing a reduction of the cross sectional area of the metalization, and an associated increase in resistance."

Editorial Note: The increase in resistance was later found to be far greater than the changes in area, but rather orders of magnitude formed due to Kirkendall voids.

In 1966, Ono and Kojima reported that the specific resistance of Al-Au compounds is not large enough to cause abnormal increase in resistance [9]. This is an important result, confirming that the changes in resistance due to morphing intermetallic compounds do not cause actual electrical failures.

William Gianelle stated in 1966 [10].

> Initially it was felt that the purple phase, $AuAl_2$ was responsible for the failure of bonds on silicon planar transistors using the gold-to-aluminum system. This Is not the case. The results of a Bell Laboratories experiment in 1962 indicate that the problem is with the tan phase Au_2Al.

Gianelle (1966) also concludes,

> Intermetallic formation can be eliminated under the bonds by controlling the amount of aluminum metallization on the chip.

Schnable, Keen and Loewenstern (1967a) wrote a report on the *Study of Contact Failures in Integrated Circuits*. They themselves list it for the year 1966 in another of their papers [11].

> "The change in sheet resistivity of the Al metallization as a result of the temperature stresses applied to these bonded structures would result in an overall resistance increase of less than 1 % of that actually observed. Accordingly, the observed resistance increases may be attributed to interaction at the Au-Al interfaces. In these tests in the temperature range from 150 °C to 210 °C, an activation energy of 24 kcal/mole [1.04 eV/atom] for degradation of these structures was … determined."

> "Thin-film studies have … been performed on the Au-Al system using the change in sheet resistivity of 500 Å Au films evaporated on 500 Å Al. Data on the changes in sheet resistivity, obtained in the temperature range of 175–275 °C, also indicated an activation energy of 24 kcal/mole."

Schnable et al. reported that actual observed resistance increase was more than ten times the sheet resistivity change.

In the same year, Schnable, Keen and Loewenstern (1967b), wrote and published their findings in *Mechanisms of Contact Failure in Semiconductor Devices* [12],

> "A study of the mechanisms of failure in ohmic and expanded contacts, including metal-semiconductor contacts and bonds to metallization in semiconductor devices, indicates that within a number of systems degradation or catastrophic failure has been produced by diffusion which produces solid solutions and/or compound formation. A great many systems are thermodynamically capable of interaction, but the important consideration is not whether they interact at all but whether the reaction proceeds at a rate which will result in metallurgical changes which adversely affect device reliability."
>
> "… the activation energy for Au-Al interaction, obtained from thin-film resistivity measurements, is in good agreement with that obtained from device life test results, and from other physical test methods, over the temperature range of 20–275 °C."

Selikson noted in 1967 that [13],

> In the aluminum-gold system, the formation of the purple plague compound and other intermetallics is catalyzed by the presence of silicon, causing rapid degradation of bond strength. The compound formation up over the oxide is accompanied by void formation as a result of [the] Kirkendall effect, leading to mechanically open bonds.

The year 1967 was particularly prolific. It was in the same year that Chen (1967) measured the change in electrical resistivity of ultrasonic gold wire bonds over a range between 156–252 °C, and found the resistance to change as bonds degraded. He observed that the Au/Al resistance increase was accompanied by a decrease in bond strength (i.e., an increase in the resistance of the Au-Al bond indicated the level of its degradation). Chen found the activation energy to be 13 kcal/mol [0.56 eV/atom] [14].

> The gold wire-aluminum thin-film joint is treated as a diffusion couple at low concentration, but above dilute levels. Electrical resistance of the joint is directly related to the solute concentration and, therefore, is a measure of the thermal annealing process… Experimental results on ultrasonically-joined gold wire-aluminum thin films showed resistance increase with decreasing bond strength. From an Arrhenius-type plot at 105 °C, 156 °C, and 203 °C, the apparent activation energy for thermal annealing of ultrasonically-joined gold wire-aluminum thin-film bonds is found to be 12.8 kcal/mole [0.56 eV/atom]. The diffusion model is independent of the method by which the two metals are joined, however.

A 1967 study by Spalvins and Buckley revealed [15],

> If a gold film is vapor deposited on an aluminum substrate at about 220 °C, a brittle purple phase forms which has been identified as $AuAl_2$ ... it is not necessary to have a liquid in order to form gold aluminum intermetallic compounds. This reaction occurs at relatively low temperatures during vapor deposition… This interface also has a narrow distance over which the transition takes place, thus being again sensitive to internal stresses and differences in the coefficient of thermal expansion between the intermetallic compound and the substrate.

Sparling summarizes the implications of purple plague in *Corrosion Prevention/Deterioration Control in Electronic Components and Assemblies* in 1967 [16].

In their 1968 work, Takei and Francombe studied interfacial strains [17],

> Mechanical separation of contacts in Si devices is known to originate in interdiffusion effects which promote the growth of intermetallic compound phases. The structural differences in these phases suggest that interfacial strains may develop between them. Using a novel beam bending approach, these strains have been measured and their variation during aging correlated with the conditions and rate of phase formation.

They observed stress values exceeding the bulk strength of Al.

The results of Francombe, Noreika and Takei's (1968) X-ray study are as follows [18],

> Comparison of the results with X-ray data for bulk samples shows that, except for Au_2Al, the structures of the thin film phases are the same as those assumed by bulk crystals of the same nominal composition.

In 1969, Anderson. Jr. and Cox studied *Failure Modes in Gold-Aluminum Thermocompression Bonds* [19]. The data shows that electrical failures occur before the mechanical failures, eliciting the role of voids in reducing the cross-sectional area of the bond.

> Electrical opens and mechanical fracture have been reported as two distinct failure modes in gold-aluminum thermocompression ball bonds. It is found that the thickness of the aluminum film to which the gold wires are bonded and the temperatures at which the bonds are held during aging determine which failure mode will be observed. The electrical failure mode predominates when bonds made to aluminum films 5,000 Å or more in thickness are aged at temperatures above 200 °C. However, at 200 °C both failure modes are observed. Mechanical fracture is expected to be the dominant failure mode when ball bonds to these 5,000–10,000 Å films are aged at temperatures below 200 °C. Bonds to thinner aluminum films (500–3,000 Å) exhibited only the electrical failure mode.

In Minoru Kashiwabara and Seizi Hattori's (1969) report on *Formation of Al-Au Intermetallic Compounds and Resistance Increase for Ultrasonic Al Wire Bonding* they wrote the following [20],

> Abnormal resistance increase related with Al-Au intermetallic compounds in the interconnection of semiconductor devices using Al wire ultrasonic bonding has been observed. The following results are obtained, together with understanding of the formation of intermetallic compounds. (1) Abnormal increase in resistance caused by the connection of the cracks leads to the breaking of electrical circuit[s]. (2) The phenomenon mentioned above varies markedly with the ratio of actual contact width of a bonded part to thickness of plated Au film, and is never found the reciprocal of this ratio is over 4 at normal storage temperature less than 350 °C.

Editorial Note: A ratio of Al wire bond width to Au thickness seems like a peculiar index of quality and adequacy. We would assume that a better ratio would be Al thickness to Au thickness. By this measure, if the ratio was very large or the opposite, very small, then intermetallic compound formation would be inhibited.

> "… the failure depends, not on the formation of Al-Au intermetallic compounds, but on the geometrical shape of the connection."

It appears that the author wants to say that the failure rate is determined more by the geometrical shape due to lateral progression rather than the diffusion rate between the two metals.

> "If thickness of Au plating is known, [then the] probability of abnormal increase in resistance can be estimated by measuring width of deformation closely related with [the] width of actual contact…"

> "... the driving force in generating and growing the cracks is volume change when the compounds are formed."

Writing in 1970, Rossiter reported on *Ambient Effects on Gold-Aluminum Bonds* [21],

> "The ambient effect observed in this study has three important results: first, it was found to have a greater effect on the lifetime of Au-Al bonds than any other parameter previously studied; second, the two major types of voiding which have been reported for many years were found to be related through ambient; third, the ambient which seems to cause this effect is oxygen where the presence of oxygen significantly extends the life of Au-Al bonds."
>
> "The mechanism for the effect of oxygen on gold-aluminum bonds is not completely understood and can only be postulated at this time. It appears that the voiding which occurs in the hermetic packages is a natural reaction for a gold aluminum bond where the aluminum from the pad rapidly diffuses into the bond by surface and grain boundary mechanisms, resulting in an increase of the intermetallic region and eventual Kirkendall voiding at the interface. When oxygen is present, the aluminum is readily oxidized, and the amount of aluminum available to enter the bond is limited by the bulk diffusion properties of aluminum. The annular voiding then results from gold diffusing out of the bond faster than aluminum enters the bond."

Editorial Note: All of that is true, but Au wires were later removed from hermetic packages because the high sealing temperatures affected the bonds to, or beyond, their end-of-life point during manufacturing. Another thing to keep in mind is that there were two types of hermetic packages in 1970,

1. A "side-brazed package" with nickel lid using Au-Sn solder at ~350 °C in forming gas (N_2 + 10 % H_2);
2. A CerDIP packages with a Pb borosilicate glass to alumina substrates and a Kovar (Fe-Ni-Co) or Alloy 42 lead frame with Ag coated bonding tips, sealed at ~425 °C.

REFERENCES

[1] J. A. Cunningham, "Expanded Contacts and Interconnexions to Monolithic Silicon Integrated Circuits," *Solid State Electronics,* vol. 8, no. 9, pp. 735–745, 1965.

[2] E. M. Ruggiero, "Gold-Aluminum Adhesion and Reaction on Semiconductor Surfaces," in *Proceedings of 4th Microelectronics Symposium*, St Louis, Mo, pp. 6B1–6B4, 1965.

[3] L. E. Colteryahn and D. D. Shaffer, "Characterization of Failure Modes in Gold-Aluminum Thermal Compression Bonds," in *IEEE WESCON International Convention,* pp. 1–6, 1965.

[4] J. Vaccaro and H. Gorton, *Reliability Physics Notebook*, RADC, Griffith NY, 1965.

[5] G. Browning, L. Colteryahn and D. Cummings, "Failure Mechanisms Associated with Thermo-Compression Bonds in Integrated Circuits," *Physics of Failure in Electronics,* vol. 4, no. June, pp. 428–446, 1966.

[6] I. A. Blech and H. Sello, "Some New Aspects of Gold-Aluminum Bonds," *Journal of Electrochemical Society,* vol. 113, pp. 1052–1054, 1966.
[7] B. Brennen, "Beryllium-Gold Alloy and Article Made Therefrom." USA Patent 3,272,625, 1 September 1966.
[8] W. M. Berger, R. S. Keen and G. L. Schnable, "Reliability Phenomena in Aluminum Metallizations on Silicon Dioxide," in *Symposium on Physics of Failure in Electronics,* pp. 1–31, 1966.
[9] K. Ono and T. Kojima, "Some Problems of Au Lead Wire in Semiconductor Devices," *Review of the Electrical Communication Laboratory, NTT,* vol. 15, no. 10, pp. 141–151, 1966.
[10] W. Gianelle, "Analysis of Seven Semiconductor Metallurgy Systems Used on Silicon Planar Transistors," *Physics of Failure in Electronics, RADC,* vol. 4, pp. 46–58, June 1966.
[11] G. L. Schnable, R. S. Keen and L. R. Loewenstern, "Study of Contact Failures in Integrated Circuits – Final Report Contract AF30(602)-4034," Philco-Ford Corp. for the Rome Air Development Center, Griffiss Air Force Base, New York, 1967a.
[12] G. L. Schnable, R. S Keen, and L. R. Loewenstern, "Mechanisms of Contact Failure in Semiconductor Devices," in *Proceedings of IEEE Sixth Annual Reliability Physics Symposium,* pp. 216–233, November 1967b.
[13] B. Selikson, "Failure Mechanisms in Integrated Circuit Interconnect Systems," *Sixth Annual Reliability Physics Symposium,* vol. Sixth, no. Nov. 1967, pp. 201–208, 1967.
[14] G. K. C. Chen, "On the Physics of Purple Plague Formation, and the Observation of Purple Plague in Ultrasonically Joined Gold-Aluminum Bonds," *IEEE Transactions on Parts, Materials and Packaging,* vol. 3, no. 4, December, pp. 149–153, 1967.
[15] T. Spalvins and D. H. Buckley, "Importance in Lubrication of Various Interface Types Formed During Vacuum Deposition of Thin Metallic Films," NASA Technical Memorandum NASA TM X-52305, Washington DC, 1967.
[16] R. H. Sparling, "Corrosion Prevention/Deterioration Control in Electronic Components and Assemblies," US. Missile Command, Redstone Arsenal, Alabama, 1967.
[17] W. J. Takei and M. H. Francombe, "Measurement of Diffusion-Induced Strains at Metal Bond Interfaces," *Solid-State Electronics,* vol. 11, pp. 205–208, 1968.
[18] M. Francombe, A. Noreika and W. Takei, "Thin Film and Bulk Structures of Phases in the System Gold-aluminum," *Thin Solid Films,* vol. 1, no. 5 (March), pp. 353–366, 1968.
[19] J. H. Anderson. Jr. and W. P. Cox, "Failure Modes in Gold-Aluminum Thermocompression Bonds," *IEEE Transactions On Reliability,* vol. 18, no. November, pp. 206–207, 1969.
[20] M. Kashiwabara and S. Hattori, "Formation of Al-Au Intermetallic Compounds and Resistance Increase for Ultrasonic Al Wire Bonding," *Review of the Electrical Communication Laboratory,* vol. 17, no. 9 (September), pp. 1001–1013, 1969.
[21] T. J. Rossiter, "Ambient Effects on Gold-Aluminum Bonds," in *Proceedings, Annual Reliability Physics Symposium*, 1970.

6 1970: First Detailed Failure Analysis and Metallurgy

As production volumes increased for consumer electronic products, the demand for higher manufacturing yields brought greater attention to the gold-aluminum interface. Great strides were made in understanding its interconnect behavior.

Philofsky, in 1970, published an excellent, detailed study on the formation and growth of the gold-aluminum intermetallic compounds [1, 2]. He states that voids form in the Au_5Al_2 phase, a relatively Au-rich phase, from which the Al diffused away. He found the activation energy for intermetallic formation to be 15.9 kcal/mol [0.69 eV/atom]. He also found that silicon substantially increased the amount of purple phase in the couples. He found that all the intermetallics were stronger than gold or aluminum.

In 1970, Lauffenburger and Myers documented [3],

> Typically bonds account for [between] 20 to over 60 per cent of all isolated failures. Gold-aluminum bonds are the worst offenders since they are subject to intermetallic compound formation in addition to process errors. Failure rates for the Au-Au and Al-Al systems are quite low and essentially independent of applied stress. Au-Al bonds exhibit failure rates only slightly higher than the other two systems at 25 °C but are highly temperature dependent. At 200 °C bond failure rate increase[s] by nearly 1-1/2 orders of magnitude.
>
> Process related bond defects can be effectively reduced, as well designed screening techniques applied to 100 per cent of devices prior to use.

In 1971, Philofsky presented design limits when using gold-aluminum bonds [4],

> "Fig. [6.1] is a plot of … approximate time to failure due to voiding in Au_5Al_2 as a function of temperature…"
>
> "As the supply of gold or aluminum diminishes (e.g. using thinner metallization or thinner gold plate), the curve moves to the left thus delaying failure at a given temperature to a longer time. The reason for this is that the growth of the Au_5Al_2 is stunted by lack of one species, and the kinetics are slowed. Moreover, with the thinner phase, distance to the phase boundary (which acts as a sink for excess vacancies) is lessened and excess vacancies can be more efficiently disposed of at the interface before they can coalesce to form voids."

DOI: 10.1201/9781003619178-6

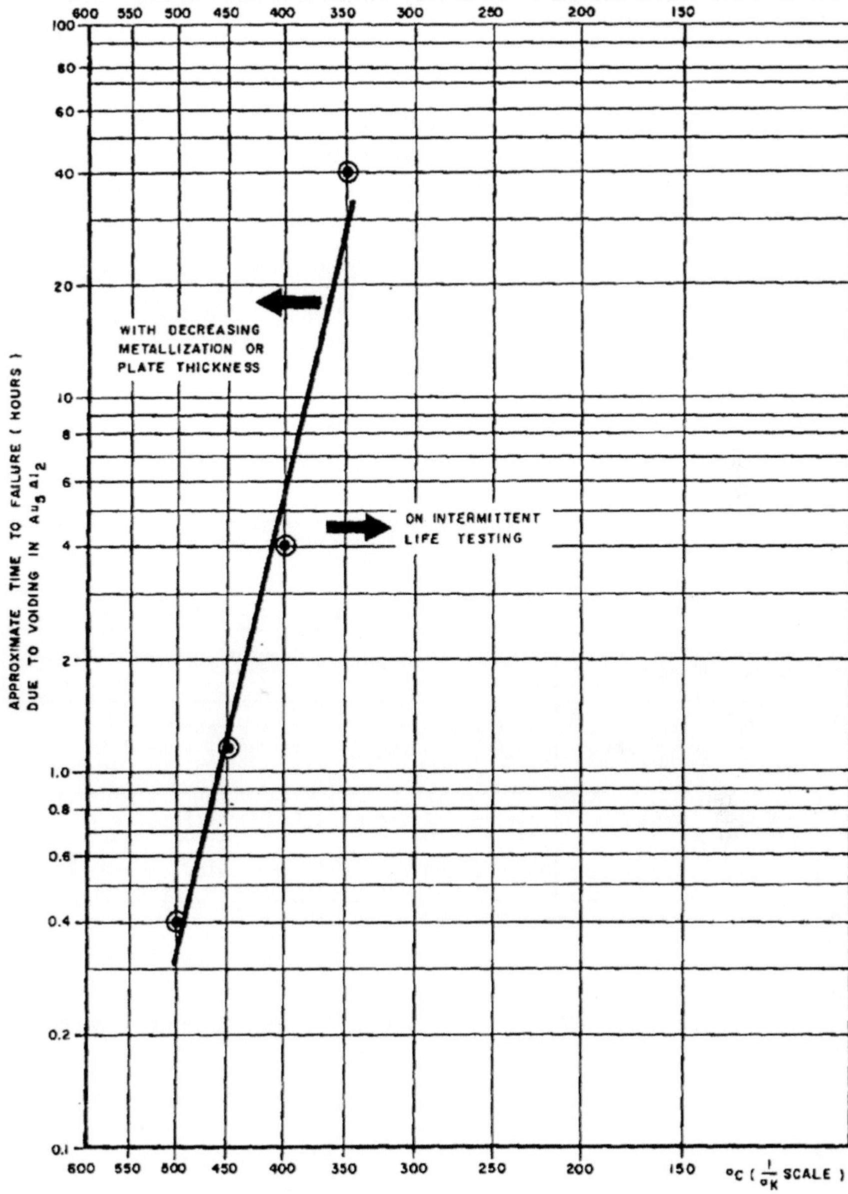

FIGURE 6.1 Time for failure caused by continuous voiding in Au_5Al_2 versus temperature.

Philofsky mentions intermetallic compounds as brittle, e.g.,

> "Microcracks occur at regions of high stress in the Au_5Al_2 due to its brittle nature…
>
> The various failure modes in gold-aluminum bonds associated with intermetallic formation can then be summarized in Tables 6.1 and 6.2. Table 6.1 can be used as a trouble-shooting guide for gold wire bonds to aluminum metallization and Table 6.2 for aluminum wire to gold plate."

He summarizes his view on design limitations in gold-aluminum wire bonds then,

> "… the following suggestions are offered to package engineers using gold-aluminum bonds.

TABLE 6.1
Failure modes associated with intermetallic formation—Gold wire to aluminum metallization

FAILURE MODES ASSOCIATED WITH INTERMETALLIC FORMATION--GOLD WIRE TO ALUMINUM METALLIZATION

	SYMPTOM	CAUSE	REMEDY
(1)	Open metallization around bonding pad	Voiding in $AuAl_2$	Keep circuit below 400°C
(2)	Zero pull strength - bond peeled off pad - fracture surface tan	Voiding in Au_5Al_2 thermal cycling will aggravate	Make metallization thinner or reduce time at temperature below those in Fig. 3
(3)	Zero pull strength - break at heel - fracture surface tan	Intermetallic formation in heel of bond fatigues during thermal cycling	Keep metallization thinner or use thicker gold wire or reduce time at temperatures below those in Fig. 6

TABLE 6.2
Failure modes associated with intermetallic formation—Aluminum wire to gold plate

FAILURE MODES ASSOCIATED WITH INTERMETALLIC FORMATION--ALUMINUM WIRE TO GOLD PLATE

	SYMPTOM	CAUSE	REMEDY
(1)	Zero pull strength - bond peeled off pad - fracture surface purple	Voiding in $AuAl_2$	Keep circuit below 400°C
(2)	Zero pull strength - bond peeled off pad - fracture surface tan	Voiding in Au_5Al_2 thermal cycling will aggravate	Make gold plate thinner or reduce time at temperature below those in Fig. 3
(3)	Zero pull strength - break at heel - fracture surface purple or tan	Intermetallic formation in heel of bond fatigues during thermal cycling	Keep gold plate thinner or use thicker aluminum wire or reduce time at temperatures below those in Fig. 7

1) The times at temperature the device encounters [i.e., device times at temperature the device encounters] should lie below the curve in Figure [6.1]. For example, a circuit containing gold-aluminum wire bonds can be aged at 400 °C for 1 hour without worry, but treating it at this temperature for 6 hours will cause some failures.
2) If the time at temperature to be encountered lies near the curve in Figure [6.1] (either above or below), the reliability of the bonds can be improved by using the thinnest gold-plate or aluminum metallization that is possible consistent with device design and the ability to bond.
3) If the device is to be subjected to much thermal cycling, the allowable time at temperature for no failures to occur will be substantially reduced from the curve in Figure [6.1]. The amount of reduction will depend on the temperature extremes, number of cycles, and quench rate.
4) If the device is to be subjected to any thermal cycling, the package should be designed so that $y > 6x$ (where y = wire thickness at heel and x = thickness of metallization) for gold-wire-to-aluminum bonds and $y > 3x$ for aluminum-wire-to-gold bonds...
5) If possible, the circuit should be designed so that all gold-aluminum wire bonds occur as aluminum wire bonded to gold plate. The reasons for this are that these wire bonds will last for longer times under identical conditions than in gold to aluminum bonds and it is also easier to specify thinner gold plate than thinner aluminum metallization.

REFERENCES

[1] E. Philofsky, "Intermetallic Formation in Gold-Aluminum Systems," *Solid-State Electronics,* vol. 13, no. 10, pp. 1391–1399, 1970a.

[2] E. Philofsky, "Purple Plague Revisited," in *8th Annual Proceedings of Reliability Physics Symposium,* pp. 177–185, 1970b.

[3] H. Lauffenburger and T. Myers, "LSI Reliability Assessment and Prediction," *Annual Symposium on Reliability,* vol. 3, no. 1, pp. 364–380, 03 Feb 1970.

[4] E. Philofsky, "Design Limits When Using Gold-Aluminum Bonds," in *9th Annual Proceedings of Reliability Physics Symposium,* pp. 114–119, 1971.

7 1971: IMC Formation Governed by Fickian Diffusion

The physics of diffusion governs the formation of gold-aluminum intermetallic compounds, their progression is affected by numerous other factors including contaminants, the bond geometry and the environment around the joint. Every kind of rendezvous in this world requires careful and thoughtful management.

In 1971, Samuel Goldfarb presented the results of a study on wire bonds applied to thick film conductors [1], using the increase in bond resistance of the aluminum bonds on varying gold thicknesses as an indicator of bond degradation. Over the course of the tests, the initial resistance values increased by a factor 5 to 7, leading to a significant reduction in bond strength.

> The tests have demonstrated a temperature induced degradation in aluminum wire ultrasonic bonds on thick film gold materials. The effect was observed at temperatures commonly employed in hybrid microcircuit processing and may be severe enough to cause spontaneous bond lifting.

In 1971, Weaver published, *Diffusion in Metallic Films* and stated that the final phase formed was Au_2Al [2].

> the time of a diffusion anneal is correspondingly affected since diffusion frequently follows a parabolic law x^2 vs. t, and because of this, diffusion can often be examined at lower temperatures than are practicable with bulk diffusion couples, without involving unduly long annealing periods.

Joshi (1971) measured the interface temperature during ultrasonic welding of a thin gold wire to a gold substrate using the thermocouple method, and the temperature was found to be no more than 120 °C [3].

Kolesnikov studied contiguous films of aluminum and gold and reported the following in 1972 [4],

> "The reaction of contiguous films of aluminum and gold is studied at temperatures between 100–300 °C with width ratios of δ_{Au}/δ_{Al} = 0.6-3.0. It was established that the character of the reaction and the composition of annealed films strongly depends on the temperature and time of exposure as well as the width ratio. The formation of intermetallic compounds already occurs with noticeable speed at 100 °C. The diffusion coefficient and diffusion activation

DOI: 10.1201/9781003619178-7

energy are evaluated by the curve of resistance change in various Au-Al width ratios. The order of magnitude of these coefficients indicates inter-crystalline diffusion."

"... the purple color of samples is related exclusively to the formation of $AuAl_2$. When compounds Au_8Al_3 and Au_4Al which are richer in gold appear, the films acquire a dark gray color. The presence of silicon is not necessary for the formation of the black product..."

The last item in Kolesnikov's study is in contradiction with Selikson (1964) [5],

> The interaction of gold and aluminum films, accompanied by the formation of intermetallic compounds, occurs even at 100 °C. With increased temperature, the rate of the processes increases greatly and the character can change, depending on the thickness ratio of the films.

In 1972, Horsting published the results of his study examining the effects of ultrasonic aluminum bonds on both (1) pure gold plated to Ni and (2) contaminated gold plated to Ni. This was done to compare the performance of pure versus contaminated gold when plated to Ni. [6].

> In the case of a good bond (pure gold), the diffusion front moves without impediment right through the gold layer and stops at the nickel interface...
>
> In the case of a bond lift failure (contaminated gold), the diffusion front starts moving into the gold also, but becomes arrested before reaching the nickel layer, while the structural failure occurs at the plane [level] where the diffusion front stops. Since we have established that the failure is connected directly with the presence of impurities in the gold plating, the following explanation may be suggested. Impurities in the gold are expelled and concentrated directly ahead of the moving diffusion front by a process similar to zone refining. When the concentration of these impurities rises above some critical level, precipitation occurs. The precipitate particles act as sinks for vacancies produced by the diffusion reaction and small voids develop, which eventually join to cause a separation. This separation acts increasingly as a diffusion barrier to the gold and eventually shuts off the gold supply to the diffusion zone, as evidenced by the fact that some of the gold plating always remains in the location where the lifting has occurred.
>
> Horsting (1972) concluded,
>
> 1. The appearance of the purple aluminum-gold intermetallic phase in an ultrasonic bond is no cause for bond failure.
>
> Editorial Note: But any purple phase at manufacture treads toward a dangerous cliff.
>
> 2. The "bondlift" type of failure, observed in this study, was shown to be caused by the presence of impurities [Ni, Fe, Co, B] in the gold.

3. The failure mechanism appears to operate through formation of a void layer, which results from impurity accumulation ahead of the Aluminum-Gold diffusion front.

Editorial Note: The statement, "ahead of the aluminum-gold diffusion front" is not consistent with Philofsky, who says that the void formation layer is within Au_5Al_2.

Bair, Herr, Poe and Fox (1972) reported the results of their study investigating the reliability of linear integrated circuits in ceramic and plastic packages [7]. They observed failures due to open gold-aluminum bond contacts and termed the failure mode as intermetallic deterioration. Concerning aluminum corrosion, they observed,

> The gold-aluminum couple at the bonds is an ideal galvanic corrosion cell in the presence of sodium chloride. The galvanic voltage between these highly dissimilar metals may exceed 3.0 volts and chloride ions promote the solubility of normally insoluble oxidized corrosion products from the anodic aluminum.

They also reported gold migration,

> Gold is usually corrosion resistant and insoluble. However, in the strong positive fields at the Pin 9 and 3 wire bonds in Unit 3014, gold solution and transfer took place with redeposit of the dissolved metal on negative cathodic surfaces such as the unpassivated edge of the substrate. This transfer of gold indicates chloride ions were probably present in the penetrating liquid films... This is the only device with gold migration observed in this program.

Concerning intermittent failures, they made an interesting observation,

> "The intermittent MTC [Monitored Temperature Cycle] sites are believed to have been "healed" during testing by a phenomenon more common in plastic encapsulated devices than in non-plastic hermetics. When a[n electrical] bias is applied across a closely spaced void, the high field may cause dielectric breakdown, or arc-over, which leaves a conductive path of carbonized plastic or rewelded metal to maintain continuity. In hermetic devices without plastic overlays aluminum runs are almost never successfully re-established by this means, while gold aluminum bonds isolated by Kirkendall void formations, may be readily healed. These bonds, however, are sometimes mechanically weak without a surrounding plastic support and can be re-opened with shock or vibration. … This self-healing property in plastic devices may be of advantage in maintaining operation in a working circuit if an open develops as a result of electro migration. However, initial detection of opens and failure analysis is difficult."
>
> "… Retention of bonds to the chip pads indicates good mechanical adhesion but not necessarily good electrical continuity."

They also observed "loss of bond adhesion at the gold-aluminum interface" after temperature cycling followed by thermal shock.

Harry A. Schafft (1972a) published a comprehensive study on wire bond electrical connections [8]. The study was titled, *Wire Bond Electrical Connections: Testing, Fabrication and degradation* and in this publication, Schafft examines the fabrication methods, materials and environmental factors that influence bond integrity.

In the same year, Schafft (1972b), published a second study, this time, a comprehensive survey of testing and fabrication of wire-bond electrical connections, explained and analyzed related materials, processes and the measurement of quality and reliability of the interconnects [9].

Horsting (1973) connected gold-aluminum bond failures with the presence of impurities in the gold and proposed a failure mechanism [10].

Following, Ramsey (1973) discussed the gold wire itself [11],

> "Gold bonding wire is cold-drawn from a purified gold ingot through diamond dies of various sizes to final end products from 0.5 to 1.5 mils diameter for typical uses in transistors and integrated circuits. Break load rather than ultimate tensile strength is most often reported in the semiconductor industry since such data [is] more easily related to processing and quality control. A relatively *soft* wire might have an *elongation* of 5 % and a break strength of 6–7 grams. A typically hard 1.0 mil gold wire will have an elongation of 1.0 % and 10–12 gms break load. Wire for bonding may be stress-relieved by various amounts to provide the desired elongation along with an acceptable strength."

> "In general, bonding wire is relatively soft with elongations in the range of 0.5 % to 15 %..."

Editorial Note: There are many annealing steps among the several wire drawing steps, so the final anneal will determine elongation.

> "The purity of the gold wire is a key factor in controlling its mechanical properties. Some gold bonding wire is sold or purchased as 99.99 % pure gold. The remaining 0.010 % (100 ppm) generally consists of Pb, Mg, Cu, Sb, Si, Ag, and occasionally other elements as the major impurities. A good deal of bonding wire is sold as 99.975 % pure which may be more realistic. Most spectrographic data indicate that wire labeled 99.99 % by the vendor has 150–200 ppm impurities. Impurities, even on this small scale, are important in controlling the final mechanical properties of the wire. Wire having a spectrographic purity of 99.999 % and a relatively large amount of residual cold-work has been observed to rapidly change physical properties after a few days storage."

> "Gas impurities are more difficult to detect and generally cause problems at the drawing stage. It is sufficient to say that if outgassing blow-holes are obtained in a bonding wire, they could definitely cause problems. Many wire manufacturers use vacuum-processed gold. Such processing should be specified by the user in any case and gas impurities as well as any resulting blow-holes would be minimized as a potential problem."
>
> Gold ball bonds are made by a thermal compression technique… Wire is threaded through a heated capillary (150–200 °C) slightly larger than the wire diameter. A ball is formed at the end of the wire by oxy-hydrogen flame cut-off melting. In addition, a weight of about 50 grams or more is used to force the ball down onto its bonding area which is heated to 300 °C or more. The effect of substrate heat, physical condition of the heated wire, and pressure, create a gold-gold or gold-aluminum metallurgically continuous bond or weld. The metallurgical characters of the wire and bond are drastically changed as they

go through the bonding process. The original structure is very fine-grained and results from the starting ingot and subsequent drawing or working processes. As the wire proceeds through the bonding process, a complete change in structure will result; i.e., large grains occur in and near the ball. Very high capillary temperatures (> 300 °C) have been observed to cause annealing of the wire and loss of initial strength. A capillary temperature of 190 °C held for five minutes did not result in modifying the original mechanical properties of the wire.

Editorial Note: Capillary temperature is expected to have a small effect relative to the initial bond temperature (300 °C) and the exothermic nature of Au-Al intermetallic compound formation. Copious intermetallic compounds are formed in a thermocompression bond.
Ramsey (1973) continues,

"Recrystallization, recovery, and grain growth are the physical processes which constitute annealing. Annealing removes the stress which has been introduced into the wire through the cold-work process before recrystallization can take place…"

"Some additives may form undesirable compounds while others form extensive or complete solid solutions. Solid solution is the most desirable situation for a useful gold bonding wire."

Editorial Note: Well said.

In the case of gold bonding wires, an addition of from 1 % to 5 % is a practical range for the alloy. The effect of this addition is to provide resistance to slip or to movement of dislocations which constitutes slip. Any alloy ingredient will distort the structure to a certain extent and provide a resistance to the movement of dislocations. The more the distortion of the basic gold structure, the greater the effect or, change in the basic properties, i.e., the greater the hardness, tensile strength and overall resistance to mechanical failure. Because of the misfit between atoms, there is also more resistance to formation of large grains and an obstruction of diffusion between grains. Alloying, therefore, tends to maintain relatively small grains through the original gold wire matrix.

Editorial Note: This is all likely true, but the relevant region is that of the intermetallic compound rather than the wire or the thin film.
Lastly, Ramsey (1973) states that,

"One problem in the use of alloy wires is the formation of teardrop shaped balls because of liquidus-solidus regions (plastic range) rather than sharp melting points observed for pure Au or eutectics. Another disadvantage is the difficulty in making bonds because of the increased hardness. This can be alleviated by annealing and increasing the elongation, but this often defeats the purpose of using a stronger wire. The effect of higher bonding or working temperatures improves the potential for relief of stress introduced during bonding and will result in an increase in ductility when compared to lower bonding temperatures for a given wire."

> "A still more discouraging problem is the apparent increased diffusion of aluminum into gold where more grain boundaries are maintained in a ball, i.e., intergranular diffusion becomes more favorable. A comparison study made of Au and Au-5 % Pd bonds shows evidence that the larger number of grain boundaries in the alloys enhances the uptake of Al and causes much quicker failure of the bond due to normal Au-Al compound formation and voiding from Kirkendall effects. The alloy wires might still give the desired effect if relatively low temperatures are maintained in processing and in use."

Editorial Note: The positive effects of impurities on bond lifetime were confirmed and verified many years later by Noolu, Murdeshwar, Ely, Lippold and Baeslack (2004), 33 years later after the original findings [12]!

Impurities can alter the reflectivity of the bonding surface. In 1973, Giorgio Riga showed [13].

> there exists a strong correlation between discoloration and bondability.

In 1974, Masami Onishi and Koichi Fukumoto studied diffusion formation of intermetallic compounds in Au-Al couples using evaporated Al films [14].

> "… the ratio of the Au_5Al_2 layer thickness to the diffusion layer thickness decreases with decrease in annealing temperature. At lower temperatures it would be expected that the diffusion layer is occupied by the Au_2Al phase layer."

> "The purple intermetallic compound $AuAl_2$ existed as an extremely thin layer and developed in all cases where cracks were observed in the Au_5Al_2 layer. Purple plague i.e., the development of the $AuAl_2$ phase layer at the gold-aluminum bonding interface of semi-conductor devices, may possibly be not a cause but a result of bond failure."

Editorial Note: Purple phase was considered as "collateral damage" and characterized as a by-product of Kirkendall voids due to asymmetric diffusion rates.

In 1974, Harman presented an overview of metallurgical failure modes in wire bonds [15].

> voids are formed by the different diffusion rates of gold and aluminum as each diffuses into the other.

Editorial Note: The lattice parameter for Al is 4.05 and 4.08 for Au. Perhaps, Al diffuses into Au more rapidly due to its smaller atomic size?

> Three types of plague-induced bond failures have been observed. In the first, the bond may be mechanically strong, but it can have a high electrical resistance or even be open-circuited. …which typically occurs with gold-wire bonded to thin aluminum metallization

Editorial Note: There are voids within intermetallic compound.

> voids form around the bond periphery limiting the available conduction paths… In the second type of failure, the voids lie beneath the bond... In this case, the bond can fail due to mechanical weakness.

TABLE 7.1
Failure Modes Associated with Intermetallic Formation (condensed from Philofsky [16])

SYMPTOM	CAUSE	REMEDY
OPEN METALLIZATION AROUND BONDING PAD (A) ZERO PULL STRENGTH - BOND PEELED OFF PAD - FRACTURE SURFACE PURPLE (B)	VOIDING IN $AuAl_2$	KEEP CIRCUIT BELOW 400°C
ZERO PULL STRENGTH - BOND PEELED OFF PAD - FRACTURE SURFACE TAN (A) AND (B)	VOIDING IN Au_5Al_2 THERMAL CYCLING WILL AGGRAVATE	MAKE METALLIZATION THINNER OR REDUCE TIME AT TEMPERATURE
ZERO PULL STRENGTH - BREAK AT HEEL, FRACTURE SURFACE-TAN (A) - TAN OR PURPLE (B)	INTERMETALLIC FORMATION IN HEEL OF BOND FATIGUES DURING THERMAL CYCLING	MAKE METALLIZATION THINNER OR USE THICKER WIRE OR REDUCE TIME AT TEMPERATURE

(A) GOLD WIRE TO ALUMINUM METALLIZATION
(B) ALUMINUM WIRE TO GOLD PLATE

The third type of bond failure occurs at the heel rather than the ball, where the metallurgy is substantially different.

Editorial Note: Philofsky is likely incorrect in claiming the presence of intermetallic compounds at the heel, which typically involves a gold wire bonded to silver plating. Since all compositions in the Au-Ag system are fully miscible, no intermetallic compounds are expected to form.

> "In addition to wire annealing, intermetallic compounds may form wherever gold and aluminum are bonded together... However, when the device is operating, [the] remaining conducting paths through annular voided regions … may be further reduced in size by electromigration, and failure could occur more quickly by this combination than by either mechanism alone..."

> "Villella [and Nowakowski (1970, 1971)] [16, 17], [who were the first to study this specific issue, conducted] extensive statistical tests with cycled devices and determined that aluminum ultrasonic bonds were more reliable in this service than were aluminum thermocompression bonds. Ravi experimentally investigated the metallurgical flexure-fatigue of a number of aluminum alloy wires and showed that aluminum -1 % magnesium wire was superior to the commonly used aluminum -1 % silicon alloy. Phillips, [Anderson and Erickson (1962)] [18] calculated wire bond geometry effects and recommended that the loop height be at least 25 % of the bond-to-bond spacing [in order] to minimize bond flexure."

> "[Philofsky (1970) states that g]old-aluminum intermetallic compounds are stronger than that of pure metals [19] providing they are void free. [Philofsky (1971) further elaborates that these gold-aluminum intermetallic compounds,

while stronger,] are also more brittle [20]. Thus, if a bond system contains plague, that system is far more susceptible to flexure fatigue than gold or aluminum wires alone."

In 1974, Rodriques de Miranda and Oswald. studied the *Changes in Strength and Resistance After Burn-In of Aluminum Wire Bonds to Thick and Thin Film Gold* [21]. Their storage at a temperature of 125 °C for 250 hours mostly resulted in a growth of various intermetallic compounds showing small changes in resistance. Their temperature-time combination was too low to lead to significant resistance increases due to void proliferation.

In 1975, Philofsky and Hall [22] summarized theirs and others' previous findings briefly in a publication titled, *A Review of the Limitations of Aluminum Thin Films on Semiconductor Devices*.

In the same year of 1975, Richard W. Vook reported his results on gold-aluminum interdiffusion phenomena [23].

> The experiments carried out during the past year showed:
>
> 1) that it was possible to identify the various Au/Al phases that form in polycrystalline films during interdiffusion at high temperatures using transmission electron diffraction techniques;
> 2) that the formation of intermetallic phases in gold-aluminum bilayers introduces severe strains in the films;
> 3) that most polycrystalline bilayer films tend to have microstructures that do not lend themselves to carrying out high resolution transmission electron microscopy in a satisfactory manner;
> 4) that epitaxial Al films can be grown at room temperature on epitaxial Au films originally formed on NaCl; and,
> 5) that such monocrystalline bilayers are much better forms of these materials to use for transmission electron diffraction work. The ultimate objective of this particular study is the evaluation of the effects of gaseous impurities on interdiffusion in gold-aluminum junctions; this is a process relating to the failure of gold-aluminum junctions in microelectronic circuits.

In 1975, Campisano, Foti, Rimini and Mayer performed activation energy calculations and presented the results as graphical data [24].

> Phase formation was studied in Au-Al thin films by means of MeV He^{+2} backscattering and glancing angle X-ray diffraction techniques. In the initial stages of compound formation where both unreacted Au and unreacted Al layers are present, the phases Au_5Al_2 and Au_2Al are found. The end phases of the low temperature treatments (150–300 °C) are $AuAl_2$ or Au_4Al in the presence of only unreacted Al or Au, respectively.
>
> The kinetics of the growth of the Al-rich compounds follow a $(time)^{1/2}$ dependence (Fickian diffusion). The activation energies for the growth of Au_2Al and $AuAl_2$ are 1.0 and 1.2 eV/atom, respectively.

Kemeny and Kalmar (1975a, 1975b) published test results of a 160 x 10^6 device-hour reliability assessment and electrical failure analysis of TTL SSI Integrated Circuits [25, 26].

> the resistivity increase is considerable as compared to the 2.3 and 3.2 μohm-cm values of pure Au and Al contacts, respectively, and can account for the increased resistance of bonds after ageing, especially by the gold-rich intermetallics, but this is only the minor cause. More important in bond deterioration is void formation in the intermetallics or in the adjacent aluminized conductor strip. Au-Al intermetallics are generally brittle.

The year 1975 was truly significant, because it was in the same year that Oswald and Rodrigues de Miranda presented changes in the strength and resistance of aluminum to gold ultrasonic bonds after temperature, electrical, and environmental stresses were applied [27].

> "... sharp increase in bond resistance above 150 °C ... appears to be primarily a temperature and time related phenomenon and is attributed to the formation of intermetallics and the appearance of Kirkendall type voids. ... annular voiding is thought to be the cause of the observed increase in bond resistance."

> "(1) hybrid circuits should not see prolonged processing, burn-in, or use temperatures in excess of 150 °C, and (2) pull strength data normally used exclusively to establish bond acceptability, by itself, can be misleading."

> "The beneficial effects of N_2 at the higher temperatures contradict the findings of Rossiter..., while the somewhat detrimental effects of the lower temperatures tend to confirm his findings."

> "Parylene application has no detrimental effect whatsoever. Rather it enhances the mechanical integrity of the bond. Bond quality after storage at 125 °C is still in excess of uncoated bonds."

In 1975, Carpenter and Fitch published *Thermomechanical Testing Techniques for Microcircuits*, in which they reported [28],

> Gold-aluminum intermetallics were formed in the pad bonds which were weakened by thermal cycling. The failure mode was "opens" between the metal stripe and the gold bonding wire due to Kirkendall voiding around the outer edge of the bond between the bond and the metal stripe. These failures often failed for intermittents first before the wire was moved enough to indicate a hard open.
>
> Section ... verifies that 1.5 mil diameter wire is more reliable than 1.0 mil wire as recommended in ... [20] and [11]. In all cases, the 1.5 mil samples outperformed the 1.0 mil diameter wire samples.

In 1976, Newsome, Oswald and Rodrigues de Miranda presented *Metallurgical Aspects of Aluminum Wire Bonds to Gold Metallization* [29].

> 1. The relatively thick gold used on ceramic substrates produces greater volumetric expansion and mechanical stress at the intermetallic interfaces of the bond when subjected to elevated temperatures. Because of the thick

gold, there is also a higher incidence of Kirkendall voiding occurring at the gold-intermetallic interface prior to the depletion of the gold under the bond.
2. Impurities included in thick film frit enhances intermetallic formation and voiding at the gold-intermetallic interface. This leads to the rapid growth of the intermetallics at relatively low temperatures.
3. Rough, porous surfaces associated with thick film inks may cause spotty bonding and a nonuniform intermetallic growth. This condition also increases the risk of voiding and cracking due to the increased stress at the interface.
4. Cracks which form around the bonds tend to isolate the conductor from the bond-and thereby increase the electrical resistance.
5. Copper as an additive to thick film inks is not as detrimental to bond integrity as frit since it tends to inhibit the formation of the intermetallics for a given temperature.

In 1976, Ram Kossowsky and Arnel I. Robinson presented *Investigation into Failures of Al Wires Bonded to Au Metallization in Microsubstrates* [30].

> "... the ultrasonic bonding creates an interaction zone about 5 μm deep. Low bond strength and failure of the bond after H_2 soldering cycle is attributed to impurities in the Au inks. These impurities form brittle intermetallics with Al and brittle, readily reducible, low melting glasses."

"1. The bond extends beyond a surface reaction and indicates interaction between the Al wire and the Au metallization.
2. Bi-rich glass nodules and Al-Pd-Pt intermetallics are responsible for a low bond strength, mainly due to embrittlement of the Au metallization Al wire reaction zone.
3. Reduction of the glass during the brazing cycle in H_2 results in a catastrophic failure of the bond.
4. Most of the impurities are diffusing from the bottom layer of Pt-Au."

In his 1976 review of the evolution and current status of aluminum metallization, Arthur J. Learn commented [31],

> With thermocompression wedge or ball bonding [32] of gold wire to aluminum, the potential exists for reduced reliability because of possible intermetallic formation and concomitant void formation [33–36]). It was proposed that silicon has the effect of accelerating such interaction and that a black ternary product so formed [5] is more detrimental than the $AuAl_2$ intermetallic which is purple in color. This effect of silicon was corroborated ([19, 37]) in another investigation. Recent work [20] has firmly established that reliable gold-aluminum bonds are also achievable.
>
> Careful attention must, however, be accorded the amounts of each material available for intermetallic formation, the bonding and postbonding temperatures and the time at temperature.

REFERENCES

[1] S. Goldfarb, "Wire Bonds on Thick Film Conductors," in *Proceedings 21st Electronic Components Conference*, pp, 295–302, 1971.
[2] C. Weaver, "Diffusion in Metallic Films," *Physics of Thin Films: Advances in Research and Development,* vol. 6, pp. 301–347, 1971.
[3] K. C. Joshi, "The Formation of Ultrasonic Bonds Between Metals," *Welding Journal,* vol. 50, p. 840, 1971.
[4] D. P. Kolesnikov, A. Andrushko and Y. Sukhinina, "Aluminum-Gold Interaction in Thin Layers," *Fizika Metallurgica,* vol. 34, no. September, pp. 529–534, 1972.
[5] P. Selikson and T. A. Longo, "A Study of Purple Plague and its Role in Integrated Circuits," *Proceedings of the IEEE,* vol. 52, no. December, pp. 1638–1641, 1964.
[6] C. W. Horsting, "Purple Plague and Gold Purity," in *10th IEEE Annual Proceedings, Reliability Physics,* pp. 155–158, 1972.
[7] B. L. Bair, E. A. Herr, A. Poe and A. Fox, "Reliability of Linear Integrated Circuits in Ceramic and Plastic Packages," Rome Air Development Center, New York, 1972.
[8] H. A. Schafft, "Wire Bond Electrical Connections: Testing, Fabrication and Degradation," NBS Technical Note 593, National Beurau of Standards, Washington, DC, 1972.
[9] H. A. Schafft, "Testing and Fabrication of Wire-Bond Electrical Connections – A Comprehensive Survey," NBS Technical Note 726, National Bureau of Standards, Washington, DC, September 1972.
[10] C. W. Horsting, "Purple Plague and Gold Purity," *Microelectronics and Reliability,* vol. 12, no. 2 (April 1973), p. 88, 1973.
[11] T. H. Ramsey, "Metallurgical Behavior of Au Wire in Thermal Compression Bonding," *Solid State Technology,* vol. 16, no. October, pp. 43–47, 1973.
[12] N. Noolu, N. Murdeshwar, K. Ely, J. Lippold and W. B. Baeslack, "Phase Transformations in Thermally Exposed Au-Al Ball Bonds," *Journal of Electronic Materials,* vol. 33, no. 4, p. 340–352, 2004.
[13] G. Riga, "TO-5 Discoloration and Bondability," *11th Annual Proceedings Reliability Physics 1973: Las Vegas, Nevada, April 3-5,* vol. 11, pp. 26–32, 1973.
[14] M. Onishi and K. Fukumoto, "Diffusion Formation of Intermetallic Compounds in Au-Al Couples by Use of Evapolated Al Films," *Journal of Japan Institute Metals,* vol. 38, pp. 148–154, 1974.
[15] G. G. Harman, "Metallurgical Failure Modes of Wire Bonds," in *12th Annual Proceedings of Reliability Physics,* p. 131, 1974.
[16] F. Villella and M. Nowakowski, "Investigation of Fatigue Problems in 1-mil Diameter Thermocompression and Ultrasonic Bonding of Aluminum Wire," NASA, 1970.
[17] F. Villella and M. Nowakowski, "Thermal Excursion can cause Bond Problems," in *9th Annual Proceedings IEEE Reliability Physics Symposium*, Las Vegas NV, 1971.
[18] R. Phillips, G. Anderson and R. Erickson, "Fundamental Failure Mechanism Studies," in *First Annual Symposium on the Physics of Failure in Electronics*, pp. 73–90, 1962.
[19] E. Philofsky, "Intermetallic Formation in Gold-Aluminum Systems," *Solid-State Electronics,* vol. 13, no. 10, pp. 1391–1399, 1970.
[20] E. Philofsky, "Design Limits When Using Gold-Aluminum Bonds," in *9th Annual Proc. Reliability Physics Symposium,* pp. 114–119, 1971.
[21] W. R. D. Miranda and R. Oswald, "Changes in Strength and Resistance After Burn-In of Aluminum Wire Bonds to Thick and Thin Film Gold," in *International Society of Hybrid Microelectronics, Proceedings of 1974*, 1974.

[22] E. Philofsky and E. L. Hall, "A Review of the Limitations of Aluminum Thin Films on Semiconductor Devices," *IEEE Transactions on Parts, Hybrids, and Packaging,* vol. 11, pp. 281–290, 1975.

[23] R. W. Vook, "Gold-Aluminum Interdiffusion Phenomena in Thin-Film Couples," Defense Technical Information Center, Rome Air Development Center (RBRT), Syracuse University, Document No. 75-2, December 1975.

[24] S. U. Campisano, G. Foti, E. Rimini, S. S. Lau and J. W. Mayer, "Kinetics of Phase Formation in Au-Al Thin Films," *Philosophical Magazine,* vol. 31, pp. 903–917, 1975.

[25] A. P. Kemeny and G. Kalmar, "Results of a 160 x 106 Device-Hour Reliability Assessment and Failure Analysis of TTL SSI Integrated Circuits, Part 1 Test Results and Electrical Failure Analysis," *Microelectronics Reliability,* vol. 14, pp. 469–498, 1975a.

[26] A. P. Kemeny, G. Kalmar and V. Stefaniay, "Results of a 160 x 106 Device-Hour Reliability Assessment and Failure Analysis of TTL SSI Integrated Circuits, Part 2 Survey of Dominant IC Failure Mechanisms and Analysis of Failure Causes," *Microelectronics Reliability,* vol. 14, pp. 499–526, 1975b.

[27] R. G. Oswald, W. R. R. D. Miranda and C. W. White, "Changes in Strength and Resistance of Aluminum to Gold Ultrasonic Bonds after Temperature, Electrical, and Environmental Stress," in *Proceedings 1975 International Microelectronics Symposium,* pp. 324–339, 1975.

[28] M. R. Carpenter and W. Fitch, "Thermomechanical Testing Techniques for Microcircuits," Rome Air Development Center, New York, May 1975.

[29] J. J. Newsome, R. G. Oswald and W. R. Rodrigues de Miranda, "Metallurgical Aspects of Aluminum Wire Bonds to Gold Metallization," in *Proceedings of Reliability Physics Symposium,* pp. 63–74, 1976.

[30] R. Kossowsky and A. I. Robinson, "Investigation into Failures of Al Wires Bonded to Au Metallization in Microsubstrates," in *Proceedings of 1976 Reliability Physics Symposium,* pp. 75–81, 1976.

[31] A. J. Learn, "Evolution and Current Status of Aluminum Metallization," *Journal of Electrochemical Society,* vol. 123, pp. 894–906, 1976.

[32] O. L. Anderson, H. Christensen and P. Andreatch, "Technique for Connecting Electrical Leads to Semiconductors," *Journal of Applied Physics,* vol. 28, no. 8 (August), p. 923, 1957.

[33] J. A. Cunningham, "Expanded Contacts and Interconnexions to Monolithic Silicon Integrated Circuits," *Solid State Electronics,* vol. 8, no. 9, pp. 735–745, 1965.

[34] G. Browning, L. Colteryahn and D. Cummings, "Failure Mechanisms Associated with Thermo-Compression Bonds in Integrated Circuits," *Physics of Failure in Electronics,* vol. 4, no. June, pp. 428–446, 1966.

[35] I. A. Blech and H. Sello, "Some New Aspects of Gold-Aluminum Bonds," *Journal of Electrochemical Society,* vol. 113, pp. 1052–1054, 1966.

[36] B. Selikson, "Void Formation Failure Mechanisms in Integrated Circuits," *Proceedings of the IEEE,* vol. 57, no. 9, pp. 1594–1598, 1969.

[37] E. Philofsky, "Purple Plague Revisited," in *8th Annual Proceedings Reliability Physics Symposium,* pp. 177–185, 1970.

8 1976: Bond Shear Era Begins

Since the days of early wire bonding experiments, wire pull testing has been a commonly used method for evaluating wire bonds. However, this technique does not always provide reliable information about the bond interface, i.e., evaluating the gold-aluminum bond using wire pulling does not offer consistent and dependable information concerning the interface. The wire may break mid-span, at the neck of the either of bonds, or cause the ball or wedge bond to lift; depending on the hook position and geometry of the wire bond. As a result, the pull values also depend on the same parameters, i.e., pull values can vary significantly based on these factors. The introduction of the ball bond shear method has made it possible to consistently obtain direct insights into the overall condition of the bond joint. Today, two key methods remain widely used for bond evaluation, namely: bond resistance measurement and shear testing. Resistance measurements are useful for detecting early changes at the bond interface, often before these changes are large enough to significantly affect or compromise the mechanical or electrical integrity of the bond.

Bushmire presented the results of his study on *Gold Aluminum Interconnect stability on Thin Film Hybrid Microcircuit Substrates* in 1976 [1].

Bushmire studied the effects of gold-aluminum intermetallic growth. Three different metallization systems were evaluated using three types of bonding wire. Tantalum-nitride-chromium-gold were bonded using 0.031 mm Al-1 % Si wire and 0.031 mm Al-1 % Mg wire. These two wire types were also bonded to substrates metallized with tantalum-nitride titanium-palladium-gold. In addition, a gold wire (0.025 mm) was bonded to Al metallization. These metallization-bond systems were evaluated by exposing the test specimens to controlled temperatures and time conditions. After exposure, the samples underwent destructive loop pull tests, destructive bond shear tests, and four-point probe electrical resistance measurements to assess their performance

The data indicates that the electrical resistance shows drastic increases long before mechanical integrity is degraded, i.e., electrical resistance begins to increase significantly well before any noticeable degradation in mechanical integrity occurs.

Thomas, Winchell, James and Scharr presented their study of *Plastic Outgassing Induced Wire Bond Failure* in 1977 [2].

DOI: 10.1201/9781003619178-8

> It is proposed that a chemical attack by halogen containing gases on the intermetallic structure initiates a growth change in the microstructure. This increases the amount of voiding within the bond zone and results in the structural weakening of the intermetallic.

Editorial Note: This is a key paper with a very clever experimental design. Outgassing from the epoxy in a TO5 (Ni) lid, affected remote wire bonds (not in contact with the epoxy).

The final technical report on the reliability factors for electronic components in a storage environment prepared by Livesay and Scheibner for U. S. Army Missile Research and Development Command in September 1977 provides an overview of studies related to the effects of intermetallic growth on Au-Al wire bonds [3].

> The long-term storage reliability of electronic components in U. S. Army missile systems is approached from the viewpoint of [the] material degradation phenomena. A survey of missile systems [was conducted] to illustrate the wide range of tactical missiles and the different electronic technologies used. Data on [global] environments [or climates], particularly temperature variations, [is] included [in the feedback] to [demonstrate] the nature [and efforts] of storage environmental stresses acting on the missile[s] and [their] electronic components. [The f]ailure processes [caused by] these environmental stresses are [analyzed and] discussed in terms of mechanical, thermal-mechanical, chemical and thermal effects. Further[more,] specific discussions are included on hermeticity, polymers, oxides and statistics.
>
> The conclusions and recommendations include several items which suggest changes in component reliability policies, and recommendations are made for future research efforts in storage reliability. [Livesay and Scheibner (1977) also include a]n extensive bibliography.

Their first conclusion is,

> The tactical missiles used by the Army will normally employ state-of-the-art electronic components as of the time the missile design is frozen. [At the time of Livesay and Scheibner's (1977) report, they wrote that,] missile systems [deployed then] involve[d] a range of electronic device technologies. [Therefore,] it [wa]s impossible to compare failure rates across the full inventory of Army missiles due to the wide range of device technologies employed. [It stands to reason that, t]here will probably never be a significant statistical basis for determining missile reliability using storage failure rate data obtained from old systems since the new technologies will generally involve very different structures. Consequently, there is a need for the continued evaluation of potential storage failure mechanisms based on advancing current technologies.

In 1977, Harman and Albers reported in their paper, *The Ultrasonic Welding Mechanism as Applied to Aluminum- and Gold-Wire Bonding in Microelectronics* [4],

> A series of experiments was carried out to determine the mechanism of gold-to gold ultrasonic bonding. These experiments, including lift-off pattern studies,

clamped-wire studies, and bond deformation versus ultrasonic vibration amplitude studies, indicate that gold ultrasonic bonding takes place primarily by means of a deformation mechanism as opposed to a heating or sliding mechanism. This is substantially the same result previously obtained from studies on the aluminum ultrasonic bonding mechanism. Further, it is shown that a deformation mechanism also holds for other forms of solid phase microelectronic bonding.

In 1977, Cohen and Cross reported on *Non-contacting Infrared Measurements of Operating Temperatures in active Semiconductor Devices* [5].

The use of infrared thermal measurements is not new, but the technique described is a unique and useful application for failure analysis. Non-contacting thermal measurement techniques are essential where, contacting sensors, such as thermocouples, cannot be attached or would significantly reduce the accuracy of measurements. To determine the effectiveness of non-contacting infrared thermal measurements as a failure analysis tool, emitter and base posts of transistors were used as measurement targets. It was hoped that information regarding temperatures would be useful in evaluating the degree of gold-aluminum intermetallic formation at the posts. Although the *problem of gold-aluminum intermetallics has been eliminated at the die through the use of monometallic systems* (aluminum metallization and aluminum wire), the gold emitter and base posts remain susceptible to the formation of intermetallics. Data obtained from collector[s], base post[s], and emitter posts may be useful not only in failure analysis but also in future derating considerations.

In 1978, Dennis F. Malik reported his study on *Monolithic Bipolar SSI/MSI Digital & Linear Integrated Circuit Analysis* [6].

"Intermetallic compound formations can occur when [a] gold wire is bonded to aluminum metallization. Some of the resulting compounds provide weak or brittle bonds, or increased contact resistance. The Kirkendall effect further weakens the bond due to the formation of voids in the aluminum. [Diffusion at b]oth room temperature and elevated temperature diffusions have been reported. Reliable gold-aluminum bonds can be made. It is necessary to minimize the total mass of aluminum available for diffusion and to keep to a minimum the cumulative time-temperature product experienced by the device in both manufacture and use."

"User surveys indicated the following as principal failure mechanisms: wire bonding, metallization corrosion, intermetallic compound formations, oxide defects, and shorts from loose conductive particles in the package."

"The time/temperature effects of intermetallic compound formation and voiding around the bond as a result of the Kirkendall effect require strict design and process control for the device. Studies have indicated that these effects are reduced significantly when the gold wire thickness at the heel is greater than six times the metallization thickness. The purity of the gold is also reported as an important factor in gold wire-aluminum bonds."

> "Screening and quality conformance tests used to detect and weed out weak gold-aluminum bonds include: temperature shock/cycle, elevated storage temperature, operating D. C. and temperature, mechanical shock, centrifuge, and bond pull tests. Temperature shocks and thermal cycling at temperatures above 200 °C have been reported to degrade good bonds."

Timothy E. Turner reported on *Hybrid Microcircuit Failure Rate Prediction* in April 1978 [7].

> To determine a failure rate contribution of interconnections, field and test data were collected from hybrid and monolithic microcircuits for which all the failures were analyzed. From this a bond failure rate was determined (bond failures/part hours x number of interconnections). … An Arrhenius curve was fit[ted] to [the] data. Since the data seemed to indicate a dramatic increase[in failure rate], above 150 °C, two continuous [Arrhenius] curves were used [below and above 150 °C] [As observed,] the sharp increase above 150 °C is probably due to the formation of gold-aluminum intermetallic compounds and [the associated voiding that occurs at elevated temperatures.]

English and Melliar-Smith reviewed the *Reliability and Failure Mechanisms of Electronic Materials* in 1978 [8].

> "During its nearly 20 years of widespread use in semiconductor manufacturing, the combination of aluminum metallization and gold wire bond has both served the industry well and caused a variety of reliability-related headaches. There has been some conversion to all-gold or all-aluminum systems, but AuA1 remains a popular combination; in some instances there appears to be no feasible alternative. Concerns about both corrosion and diffusion or intermetallic phase formation continue to the present day, despite extensive study and a fairly good understanding of the basic metallurgical phenomena."
>
> "One of the alloy phases, Al_2Au, is purple, which accounts for the originally reported appearance of the bond area and the appellation "purple plague." Additional alloy phases have been identified in bulk diffusion couples, and it has been thought that brittleness of one or more of these phases accounted for bond weakening. There is no clear evidence in support of this idea, and a more likely source of bond embrittlement is the development of an array of Kirkendall voids under the bond in the course of interdiffusion."
>
> "As operating temperatures rise, it becomes more urgent to assure that plague formation will not interrupt device functioning within a normal service lifetime."

Shih and Ficalora, (1978) published *the Reduction of Au-Al Intermetallic Formation and Electromigration in Hydrogen Environments*, and wrote the following [9].

> "SEM examination showed formation of Au-Al intermetallics in the specimens treated in Argon. Specimens treated in hydrogen showed little or no compound formation."

"… observations of the effect of hydrogen on the formation of aluminum-gold intermetallics and on electromigration in aluminum are both consistent with hydrogen reducing the mass transport of aluminum."

Editorial Note: It's interesting that hydrogen has an effect, but it's not practically relevant since hydrogen is not present in packaged devices under normal conditions.

In 1978, Lundgren and Ladd reported [10],

A two-phase study was performed to investigate basic failure mechanisms, alternate fabrication methods, and operational life times of low-noise microwave MESFETs fabricated on GaAs channel layers grown by liquid-phase epitaxy. The basic sub-elements of the FET were investigated, and a Cr link for connecting the Al gate to Au bond pad was developed and incorporated into an improved FET.

The addition of Cr essentially served as a barrier between Al and Au to prevent Au-Al intermetallic formation.

At 270 °C the primary failure mode is catastrophic gate failure caused by Au-Al intermetallic formations.

In 1979, Da-Yuan Shih and Peter J. Ficalora reported [11],

"… substrates of silicon, thermally oxidized silicon, nitrided silicon, and sapphire were coated with aluminum films (4200 Å thick) across one-half [of] the surface. Gold films (8200 Å) were deposited on these same substrates such that there was a 2-mm gold overlap on the aluminum. These specimens were then heated (350 or 400 °C) for 1-3 h, in half atmospheres of H_2, Ar, O_2, N_2, He, or air and examined in an SEM. If we ignore the formation of aluminum oxide in the air and oxygen environments, … the Au-Al compounds readily formed in all environments except hydrogen. In the worst case for hydrogen the compounds formed only at the gold-aluminum contact line."

"Since hydrogen does not interact with gold and since the compounds form via a diffusion mechanism, it seemed reasonable to assume that the observed effects were due to a reduction in the aluminum flux due to an adsorption of hydrogen on aluminum grain boundaries and vacancies."

Gallace, Jacobus, Pfiffner and West compared the performance of plastic packages with hermetic packages in June 1979 [12].

Majni, Ottaviani and Galli reported on *AuAl Compound Formation by Thin Film Interactions* in 1979 [13].

Feinstein reported on the query *Do the flame retardants affect the reliability of molded plastic packages?* in 1980 [14].

"Epoxy based thermoset materials are widely used for molding of microelectronic devices. Accelerated aging of molded devices has led to increased failure rates when the molding compound contained a brominated flame retardant (FR). In this paper, we investigate the effect of a brominated FR on the failure rate of a device aged at elevated temperatures under [electrical] bias. We also ask whether aging characteristics in the 200 °C range can be extrapolated through the glass transition (T_g) to service temperatures."

"We chose an electrical/electronic, rather than semiconductor, grade novolac epoxy which previously exhibited the poorest bias aging results of a group of FR and non-FR molding compounds. Half FR and non-FR versions of the epoxy in question were also obtained for the present experiments. The test vehicle was a molded DIP with a TTL device and Tape Automated Bonding (TAB) technology. Devices were prepared with and without a silicone rubber junction coating. Devices were aged for times up to 10,000 h[ours] at 125, 150, and 200 °C, with and without bias. The results from time to failure and failure mode analyses were complemented by X-ray fluorescence, and aqueous extraction performed on the cured molding compounds."

"Failure rate at elevated temperature was found to be independent of the presence of FR. This result is consistent with our finding of four times as much extractable Cl– as Br–, and with the identification of CH_3Cl as a major thermal degradation product below 200 °C, in a companion study by Lum and Feinstein. The silicone rubber junction coating retarded [i.e., slowed] failures by a factor of 5–10. Both electrolytic (operating bias) and chemical (no bias) corrosion mechanisms were operative. However, the failure rate was a factor of 5–10 lower for unbiased devices. Direct attack of the Cu beams did not appear to be a problem in this study, although leads may act as fuses after shorting of the chip metallization. Although a linear extrapolation of failure rates from 200 °C to lower temperatures is not valid, the result will be conservative."

Wagendristel, Schurz, Ehrmann-Falkenau and Bangert conducted *an X-ray Optical Study of Layered Phase Growth in Au-Al Thin Film Couples,* and reported their findings in September 1980 [15],

"The growth of Au-Al compounds proceeds markedly already below 100 °C and causes an increase of the electrical resistance and an increase of the internal stress in the bonding region."

"... Au-Al thin film couples of a few tens of nm in thickness (10 nm corresponds to ~50 atomic layers) interdiffuse at temperatures below 100 °C by the growth of a Au_2Al layer. This is in agreement with the mixing behavior in thicker film couples where Au_2Al was found to be the only growing phase. Au_5Al_2, however, which was reported to appear before Au_2Al formation starts, might be present in the amount of some atomic layers. Since its concentration differs not very much from that of Au_2Al, the existence of a small amount is below the limit of resolution of the method we used. The activation energy of the phase growth for Au_2Al formation Q = 1.05 eV/atom agrees excellently with values of 1.0, 1.02, and 1.03 eV reported [by Takei 1968, Campisano 1975 and Weaver 1971] respectively. This agreement gives further evidence that Au_2Al is the only growing phase since the activation energy for the formation Au_5Al_2, is much lower."

Editorial Note: According to Auger analysis, thallium tends to accumulate at the electroplated surface. To provide context, Auger analysis or Auger Electron

Spectroscopy (AES), is used to determine the elemental composition of a material's surface.

James (1980) studied the effect of impurities on the gold wire [16].

The following summarizes the supporting causal hypotheses.

1) Thallium is demonstrably transferred from the lead frame to the gold wire after stitch bond contact; highly mobile thallium forms a gold-thallium eutectic resulting in weakened grain boundaries.
2) Hydrogen flame-off provides a larger molten zone and may reduce impurity oxides to form trapped insoluble water which nucleates, grows, and weakens grain boundaries.
3) The metal capillary transfers heat more efficiently along the gold wire during recrystallization, expanding the range over which thermal-induced failure mechanisms can operate.
4) Copper and beryllium may form hard, but brittle, intermetallics at grain boundaries, counteracting the beneficial beryllium-gold intermetallics.

Majni, Nobili, Ottaviani, Costato and Galli studied phase formation in gold-aluminum thin film interactions, using backscattering and X-ray diffraction techniques, eventually reporting the results in 1981 [17].

Phase formation has been studied in Au-Al thin-film interactions by means of $_4He^{+2}$ MeV backscattering and x-ray diffraction techniques. The films were annealed at temperatures low enough for no liquid phase formation. Au_4Al, Au_5Al_2, Au_2Al, AuAl, and $AuAl_2$ were detected from samples prepared by depositing the exact stoichiometry quantities of Au and Al. The phase formation sequence was determined. The first phase formed is Au_5Al_2; the others are observed with the disappearance of one phase previously formed. For instance, Au_4Al grows only when all Al is reacted and, on the other hand, $AuAl_2$ when all Au is reacted. The sequence of the phases in the Au-AI interactions cannot be interpreted on the basis of pure thermodynamic or kinetic arguments and a composite mechanism is proposed.

Using ion beam microtexturing, Rossnagel and Robinson (1982) determined an activation energy of 1.04 eV/atom for Au surface diffusion into Al substrate [18].

REFERENCES

[1] D. Bushmire, "Gold Aluminum Interconnect Stability on Thin Film Hybrid Microcircuit Substrates," in *IRPS*, Las Vegas NV, pp. 55–62, 1976.

[2] R. E. Thomas, V. Winchell, H. K. James and T. Scharr, "Plastic Outgassing Induced Wire Bond Failure," in *Proceedings of 26th Electronic Components Conference*, pp. 182–187, 1977.

[3] B. R. Livesay and E. J. Scheibner, "*Reliability Factors for Electronic Components in a Storage Environment*," U. S. Army Missile Research and Development Command, Redstone Arsenal, Alabama, September 1977.

[4] G. G. Harman and J. Albers, "The Ultrasonic Welding Mechanism as Applied to Aluminum- and Gold-Wire Bonding in Microelectronics," *IEEE Transactions on Parts Hybrids Packaging,* vol. 13, no. 4 (December), p. 406, 1977.
[5] M. Cohen and E. Cross, " Non-contacting Infrared Measurements of Operating Temperatures in Active Semiconductor Devices," in *ATFA*, Los Angeles, CA, USA, 1977.
[6] D. F. Malik, "*Monolithic Bipolar SSI/MSI Digital & Linear Integrated Circuit Analysis*," Raytheon Company, Equipment Division, Life Cycle Analysis Department, Huntsville, Alabama, January 1978.
[7] T. E. Turner, "*Hybrid Microcircuit Failure Rate Prediction*," IITRI/Reliability Analysis Center Rome Air Development Center Air Force Systems Command, Griffiss Air Force Base, New York, April 1978.
[8] A. T. English and C. M. Melliar-Smith, "Reliability and Failure Mechanisms of Electronic Materials," *Annual Review Materials Science,* vol. 8, pp. 459–95, 1978.
[9] D. Y. Shin and P. J. Ficalora, "The Reduction of Au-Al Intermetallic Formation and Electromigration in Hydrogen Environments," in *16th Annual Reliability Physics Symposium,* pp. 268–272, 1978.
[10] R. E. Lundgren and G. O. Ladd, "Reliability Study of Microwave GaAs Field-Effect Transistors," in *16th Annual Reliability Physics Symposium, 1978*, April 1978.
[11] D.-Y. Shih and P. J. Ficalora, "The Reduction of Au-Al Intermetallic Formation and Electromigration in Hydrogen Environments," *IEEE Transactions on Electron Devices,* vol. ED-26, no. 1 (January), pp. 27–34, 1979.
[12] L. J. Gallace, L. A. Jacobus, E. J. Pfiffner and C. A. West, "Accelerated Reliability Evaluation of Trimetal Integrated Circuit Chips in Plastic Packages," *IEEE Transactions on Components, Hybrids, and Manufacturing Technology,* vol. 2, no. 2 (June), pp. 172–179, 1979.
[13] G. Majni, G. Ottaviani and E. Galli, "AuAl Compound Formation by Thin Film Interactions," *Journal of Crystal Growth,* vol. 47, pp. 583–588, 1979.
[14] L. G. Feinstein, "Do the flame retardants affect the reliability of molded plastic packages?" *Proceedings of the 30th Electronic Components Conference*, April 28–30, 1980, San Francisco CA. pp. 106–112.
[15] A. Wagendristel, H. Schurz, E. Ehrmann-Falkenau and H. Bangert, "An X-ray Optical Study of Layered Phase Growth in Au-Al Thin Film Couples," *Journal of Applied Physics,* vol. 51, no. 9 (September), pp. 4808–4812, 1980.
[16] H. K. James, "Resolution of the Gold Wire Grain Growth, Failure Mechanism in Plastic Encapsulated Microelectronic Devices," *IEEE Transactions on Components, Hybrids, and Manufacturing Technology,* vol CHMT-3, no. 3, pp. 370–374, 1980.
[17] G. Majni, C. Nobili, G. Ottaviani, M. Costato and E. Galli, "Gold-aluminum Thin-film Interactions and Compound Formation," *Journal of Applied Physics,* vol. 52, pp. 4047–4054, 1981.
[18] S. M. Rossnagel and R. S. Robinson, "Surface Diffusion Activation Energy Determination Using Ion Beam Microtexturing," *Journal of Vacuum Science & Technology,* vol. 20, no. February, pp. 195–198, 1982.

9 1982: Thermosonic Supplants Thermocompression – Active Intermetallic Compound Suppression

The mechanical setup for forming gold-aluminum bonds has continued to evolve over time. This advancement in technology has been primarily focused on the following: increasing throughput, improving the initial quality of the bonds, enhancing long-term reliability of devices and accommodating the ever-decreasing dimensions of interconnections (i.e., bonding machinery is being improved to handle the increasingly smaller scale of electrical connections in modern electronic devices).

Thermosonic bonds are far superior to thermocompression bonding, as Si chip temperature is typically ~150 °C rather than 300 °C for the former, so the thermosonic bond has many fewer intermetallics at time of manufacture.

In 1982, Charles, Romenesko, Wagner, Benson, and Uy presented their study on *the Influence of Contamination on Aluminum-Gold Intermetallics* [1].

> A low temperature form of rapid aluminum-gold intermetallic growth has been encountered in hybrids fabricated with aluminum thin film metallization (over a nickel-chromium adhesion layer) and the simultaneous use and curing of two epoxies (Ablefilm 517 and Ablefilm 550). Similar hybrids without the introduction of the Ablefilm 550 epoxy did not experience the same failure modes. By all surface analysis methods, the hybrids were relatively clean and did not suffer from any halogen problem which has led to other intermetallic failures in recent times. The use of Ablefilm 550 did increase the amount of silicon present on both substrate metalization and chip bonding pads prior to wire bonding (a possible source of the silicon is the silica additive in Ablefilm 550). Silicon has been shown to be an important factor in previous intermetallic growth studies. The other trace impurities found by the various surface analysis techniques could not be [definitely] correlated with the failure, although the increased calcium, chlorine and sodium signals from the Ablefilm 550 (by SIMS analysis) when compared to the Ablefilm 517 does indicate some cause for concern and further study.

In 1982, Stafford published the results of their study on the *Reliability Implications of Destructive Gold Wire Bond Pull and Ball Bond Shear Testing* [2].

> "A combination of destructive wire pull and ball bond testing is suggested as yielding the most reliable type of wire bond."

DOI: 10.1201/9781003619178-9

"... the contact surface in thermosonic gold ball bonding is divided into three regions. No bond is formed in the outer ring. The bond is formed in the inner ring and the central region but at different elevations from the "contact" plane. For a 0.001 in. diameter gold wire, the bonded area of the ball bond has a radius of 30 μm, or approximately 0.0012 in."

In 1982, Ebel, Jeffery and Farrell published *Wirebonding Reliability Techniques and Analysis* [3].

Activation energies below 1.0 eV for [the] gold/aluminum wirebonding process indicates the probability of surface contamination present in the bonding area. Activation energies of about 0.4 eV are typical of contamination caused by residues deposited on the surface during the curing of epoxies.

In 1982, Graves and Gurany presented, *Reliability Effects of Fluorine Contamination on Aluminum Bonding Pads on Semiconductor Chips* [4].

"The majority of the technical support for the previous statements has concentrated upon the study of non-hermetically sealed plastic encapsulated devices and of chlorine as the halogen of concern."

"This paper reports the occurrence of a similar problem during manufacture of hermetically sealed hybrids and identifies fluorine as the ionic contaminant, responsible for the aluminum corrosion."

"The effects of aluminum corrosion have been observed at Bendix during the fabrication of HI-REL hybrids for the Patriot missile fuze. Thermosonically formed ball bonds between gold wires and aluminum bonding pads on semiconductor chips in hermetically sealed hybrids failed after 1,000 hours at + 125 °C under electrical bias. Prior to sealing, all of the wire bonds had been subjected to and successfully passed a 3 gram non-destructive pull test."

In 1982, Plunkett and Dalporto presented *Low Temperature Void Formation in Gold-Aluminum Contacts* [5].

"Void formation in both aluminum to gold and gold to aluminum wirebonds ha[s] been [widely] investigated. Voids may form in either type of bond [and] at the same approximate location: between the gold-rich intermetallic phase and the gold of either the substrate or the ball bond. The boundary conditions are different for each type of bond such that different considerations must be given to avoid this void formation. These considerations are summarized below:

1. Void formation in gold to aluminum wirebonds may be avoided if the correct thickness of gold is used for a particular application or if alloying elements are added to the gold to inhibit the intermetallic growth.
2. Gold to aluminum wirebonds should not be qualified by accelerated aging since the void formation is dependent on both the gold thickness and aging temperature. However, early high temperature exposure of the wirebonds may actually reduce the chance of void formation at standard temperatures.

3. The greatest effect on void formation in gold to aluminum wirebonds is the amount of contamination present on the bond pad prior to making the bond. level of contamination on the bond pad before bonding.
4. Accelerated aging of gold to aluminum wirebonds is a valid qualification procedure and is a very useful procedure for determining the effects of contamination on void formation."

In 1983, Blish II and Parobek presented *Wire Bond Integrity Test Chip,* showing [6],

> "… a 20 mΩ resistance increase corresponds to unacceptable performance in bond pull and ball shear.

This method was used for optimization of gold wire bonding to pure aluminum and silicon-doped aluminum and for encapsulant selection.

In 1983, Tu, Ottaviani, Gosele, and Foll published a phase transformation analysis of *Intermetallic Compound Formation in Thin-Film and in Bulk Samples of the Ni-Si Binary System* [7].

> In this study, we attempt to clarify the difference between interdiffusion in thin-film and in bulk samples, and we have chosen the Ni-Si binary system for experimentation. A major difference is in the behavior of intermetallic compound formation; while the compounds form sequentially in thin films, they tend to appear simultaneously in bulk samples. The essence of sequential formation is the absence of the other stable phases during the growth of a specific one. We have used cross-sectional lattice imaging of the interface between a NiSi film and a Si to confirm the absence of $NiSi_2$ at the interface.

In 1984, Derby and Wallach published *Development of Theoretical Model Diffusion Bonding* [8].

> In previous work[s], a theoretical model for solid state diffusion bonding was described. Possible diffusion bonding mechanisms were identified and mass transport rate equations for each proposed in terms of both process variables, [namely, conditions] (time, temperature, pressure) and material properties. However, the mechanism of mass transfer in the vapour phase was not [been] described since, for many diffusion bonding applications, the contribution from this mechanism will not be significant. For completeness, the vapour phase mass transport rate equations now are derived. In addition, a revised model is presented for the power law creep mechanism, based on considerations of the elastic and plastic deformation of a long triangular ridge. This new approach eliminates the assumptions, implicit in the earlier work, which break down in the later stages of diffusion bonding when the interface is substantially bonded; better agreement with experimental data also is obtained.

In 1984, Gale presented *Epoxy Degradation Induced Au-Al Intermetallic Void Formation in Plastic Encapsulated MOS Memories* [9].

> High temperature aging of devices with Au-Al bonds molded in epoxy can degrade the polymer and cause failure of the bonds. The apparent activation

> energy of the bond degradation (0.8±0.1 eV/atom) is within experimental error of that for pure Au-Al intermetallic formation (0.7 eV) and that measured for the breakdown of the flame retardant (0.8 eV). Cl impurities that are bound in the epoxy resin are released; the flame retardant is activated and Br^- is formed. The gross void formation that causes the Au-Al bond failure may result from Al removal in the form of volatile halides. Further purification of the starting resins could improve the high temperature performance of epoxy molding compounds.

Editorial Note: It was later learned that the issue was Br, rather than Cl.

Gerling, published *Electrical and Physical Characterization of Gold-Ball Bonds on Aluminum Layers* in 1984 [10].

> The temporal sequence of the alternative modes of contact failure would be:
>
> - void layer in the ball on the Au-rich side of Au-Al-phases with no or little visible intermetallics around the ball causing bond-lift-off as early failures (this mode often is attributed to contamination);
> - annular voids in the converted Al-layer within Al-rich phases under the ball edge, moderate spread of Au-Al-intermetallics will be visible; bond pull strength has decreased;
> - peripheral voids due to depletion of aluminum as final wear out mechanism.
>
> Severe voiding due to any of these modes can be indicated by contact resistance measurements. Test conditions can be defined to test for contact integrity affected by final peripheral voids only. Storage at 200 °C for 2,000 h[ours] and allowable increase of resistance $\Delta R \leq 1$ ohm is a useful test condition to assess bond life time.

Gerling identified three different phases for gold-aluminum contact resistance change during the lifetime of a wire bond and related them to the growth of Au-Al-phases.

The first phase was attributed to the formation of intermetallic compounds during a short time, resulting in slight resistance increase. The second phase was a slow increase of contact resistance due to the spread of porous Au-Al-compounds over a long period. The final phase was a fast increase of contact resistance due to voiding near the periphery of the Au-Al-compound layer. Gerling (1984) considered $\Delta R \leq 1$ ohm as "a useful test condition to assess bond life time."

In 1984, Clatterbaugh, Weiner and Charles presented their study on *Gold-Aluminum Intermetallics: Ball Bond Shear Testing and Thin Film Reaction Couples* [11].

> "… the shear failure mechanisms for thermosonic gold ball aluminum thin film bonds are different in origin for Si and SiO_2 substrates. … neither contact system is significantly superior to the other after prolonged thermal aging."
>
> "… the shear strength is proportional to the weld circumference (as opposed to the weld area) and the aluminum metallization thickness."
>
> "For the case of the Au-Al-Si contact system, the primary failure mode for prolonged high temperature annealing appears to be the continuous formation of gold-rich intermetallic phases."

In 1984, John R. Devaney and Roger M. Devaney presented *Thermalsonic Ball Bond Evaluation by a Bond Pluck Test* [12].

> The thermosonic (thermalsonic) ball bonding process, because of the low temperatures involved, often results in poor bonds due to spotty intermetallic formation. The typical destructive or non-destructive pull tests are inadequate to detect this condition. By the use of a simple pluck test which removes the ball, the footprint interface can be visually examined. A reaction across 50 % or more of the interface is indicative of a satisfactory bond. By a simple, quick, routine test at the bonding machine, if necessary, the extent of true bond formation can be determined.

In 1985, Tom D. Hund and Paul V. Plunkett published their study on *Improving Thermosonic Gold Ball Bond Reliability* [13].

> several orders of magnitude increase in reliability can be achieved from using optimum wire bond parameters. The ball shear force value has also been shown to be very useful in predicting reliability and thus eliminating time consuming wire pull tests.

REFERENCES

[1] H. K. Charles, B. M. Romenesko, G. D. Wagner, R. C. Benson, and O. M. Uy, "The Influence of Contamination on Aluminum-Gold Intermetallics," in *IEEE/PROC. IRPS*, pp. 128–139, 1982.

[2] J. W. Stafford, ". "Reliability Implications of Destructive Gold Wire Bond Pull and Ball Bond Shear Testing," *Semiconductor International,* vol. 5, no. May, p. 83, 1982.

[3] G. H. Ebel, J. Jeffery and J. P. Farrell, "Wirebonding Reliability Techniques and Analysis," *IEEE Transactions on Components and Manufacturing Technology,* vol. CHMT-5, no. 4. December 1982, pp. 441–445, 1982.

[4] J. F. Graves and W. Gurany, "Reliability Effects of Fluorine Contamination on Aluminum Bonding Pads on Semiconductor Chips," in *32nd Electronic Components Conference*, San Diego, pp. 266–267, 1982.

[5] P. V. Plunkett and J. F. Dalporto, "Low Temperature Void Formation in Gold-Aluminum Contacts," in *Proceedings 32nd Electronic Components Conference*, pp, 421–427,1982.

[6] R. C. Blish II and L. Parobek, "Wire Bond Integrity Test Chip," in *21st Annual Proceedings International Reliability Physics Symposium,* pp. 142–147, 1983.

[7] K. N. Tu, G. Ottaviani, U. Gosele and H. Foll, "Intermetallic Compound Formation in Thin-Film and in Bulk Samples of the Ni-Si Binary System," *Journal of Applied Physics,* vol. 54, p. 758, 1983.

[8] B. Derby and E. R. Wallach, "Diffusion Bonding: Development of Theoretical Model," *Metal Science,* vol. 18, no. 9 (September), pp. 427–431, 1984.

[9] R. J. Gale, "Epoxy Degradation Induced Au-Al Intermetallic Void Formation in Plastic Encapsulated MOS Memories," in *22nd Annual Proceedings International Reliability Physics Symposium,* pp. 37–47, 1984.

[10] W. Gerling, "Electrical and Physical Characterization of Gold-Ball Bonds on Aluminum Layers," in *Proceedings IEEE Electronic Components Conference,* pp. 13–20, 1984.

[11] G. V. Clatterbaugh, J. A. Weiner and H. K. Charles, "Gold-Aluminum Intermetallics: Ball Bond Shear Testing and Thin Film Reaction Couples," in *Proceedings IEEE Electronic Components Conference,* pp. 21–30, 1984.

[12] J. R. Devaney and R. M. Devaney, "Thermalsonic Ball Bond Evaluation by a Bond Pluck Test," in *Proceedings ISTFA 1984 22-24 October 1984*, Los Angeles, California, October 1984, pp. 237–242.

[13] T. D. Hund and P. V. Plunkett, "Improving Thermosonic Gold Ball Bond Reliability," *IEEE Transactions On Components, Hybrids, and Manufacturing Technology,* vol. 8, no. 4 (December 1985), pp. 446–456, 1985.

10 1986: Take a Bromide?

Editorial Note: Humor is delightful.

The development and popularity of personal computers demanded cheaper components to produce affordable systems. Replacement of robust hermetic packages based on expensive packaging materials with plastic encapsulation required a closer look at the ingredients of the molding compounds for increased throughput, enhanced quality and long-term reliability. Increased reliability of electronic components and their affordability helped expand the use of electronics everywhere on the earth and in space.

In 1986, Syed Sajid Ahmad, Richard C. Blish II, Timothy J. Corbett, Jerrold L. King, and C. Glenn Shirley published the *Effect of Bromine in Molding Compounds on Gold-Aluminum Bonds* [1].

> while bromine accelerates the degradation of the aluminum-gold interface, the detailed mechanism can vary with the molding compound.

The authors also concluded that the observed exponential time dependence of the resistance increase could not be interpreted in terms of a simple mechanism (i.e., the rapid increase in resistance over time could not be explained by a simple mechanism). The data did not fit a $t^{1/2}$ law expected for a diffusion-controlled reaction. Indeed, log-log plots of the data proved that no power law of any kind would fit the data. It was not surprising that the data did not correspond to any simple model because the geometry and the morphology of the reaction zone were and remain complicated.

This study proved pivotal in subsequent research aimed at improving the reliability of plastic encapsulated devices, ultimately enabling future handheld devices to operate reliably for years.

In 1986, Fouracre published *Electron Microscope Observations of Chemical Diffusion in the Al/Au System* [2].

> An electron microscope technique was employed to observe … phase boundary migration in Al/Au thin film diffusion couples [directly]. There was no direct evidence of any effect of enhanced grain boundary diffusion, either in the terminal aluminum phase or in the intermediate Au_2Al phase. This phase was the only one of the five possible phases which was observed in the diffusion couple. The Au_2Al formed showed an ordering effect with time.

DOI: 10.1201/9781003619178-10

In 2004, Noolu, Murdeshwar, Ely, Lippold and Baeslack III wrote two articles on *Partial Diffusion Reactions and the Associated Volume Changes in Thermally Exposed Au-Al Ball Bonds* and *Phase Transformations in Thermally Exposed Au-Al Ball Bonds* [3] [4].

In 1986, Tim Koch, Wayne Richling, John Whitlock and Dave Hall presented *a Bond Failure Mechanism* [5].

> "Intermittent continuity failures of P-DIP parts were found to be the result of wire bond failures induced by stresses associated with the assembly process. The bond failures were characterized by cracking of the underlying pad structure. It was found that silicon nodule precipitates from the aluminum metallization in the pad acted as points of high stress during the bonding of gold to the pad regions. By altering the bonding parameters and the underlying pad structure, failure of the bonds could be prevented."

> "The bond failure mechanism found on plastic DIP parts was a result of interactions between the materials in the bond pad of the integrated circuit and the packaging stresses of the bonding and molding processes. The silicon precipitates for the Al-Si (1 %) metal film were points of high stress during the bonding operation which caused microcracks in the underlying oxide and silicon layers. These cracks were propagated during the molding operation and reliability testing of the parts resulting in an electrical failure of the bond. The solutions to the bond failure mechanism included changing or optimizing all the interacting processes. The major influence of bonding stresses was significantly reduced by modifying the bonding parameters. Removal of the deposited PSG layer from the pad structure decreased the susceptibility of the pad to cracking induced by the silicon nodules during the bonding operation. The failure via the silicon precipitates could be eliminated by removing the silicon from the aluminum film, but this solution cannot be used due to contact spiking of Al which occurs without excess silicon present in the metal layer. Finally, the molding stresses were reduced through studies of the position and number of bond failures and by basic analysis of the molding operation with respect to the flow behavior and material characterization of the plastic."

Editorial Note: Contact spiking of Al into Si was solved later by using refractory plugs to isolate Al from Si, so doping the Al with Si was no longer necessary.

In 1986, Muhib M. Khan and Homi H. Fatemi presented *Gold-Aluminum Bond Failure Induced by Halogenated Additives in Epoxy Molding Compounds* [6].

> High temperature failures of Au-Al bonds have been found to be dominated by free bromide ions present in the molding compound rather than the organic bromide species. The rate limiting step was the diffusion of bromide ions out of the resin matrix. The activation energy of failure (0.3–0.5 eV/atom) was far less than observed for either thermal decomposition (1.3 eV) or hydrolytic bromide extraction (1.6–1.8 eV). It is, however, similar to diffusive free bromide extraction in water (0.3–0.4 eV). In the total absence of free bromide ions, failure could still occur by the interaction of Al of AuAl intermetallic

with HBr released from the resin, though at a much slower rate. This, however, remains to be resolved through further studies. Meanwhile, minimization of free bromide ions in the commercial molding compound formulation should be the course of action in order to improve device reliability.

In 1986, Footner Richards, Stephens and Amos presented *a Study of Gold Ball Bond Intermetallic Formation in PEDs using Infra-Red Microscopy* [7].

A novel technique, using infra-red microscopy in the reflection mode, has been used to examine gold-aluminum bond interconnections rapidly by imaging through the bulk silicon. Its use has led to the characterization of the Au-Al interface and the degree of intermetallic formation in terms of a unique IR footprint. The database of footprints obtained from a range of heat-treated commercial product has allowed a good understanding of the bond quality and its possible effect on device reliability. Comparison with results from conventional failure analysis (FA) techniques has demonstrated its use for field failures. The method is eminently suitable as a tool for both qualification and testing exercises and for field failures.

In 1987, Gustapsson and Lindborg presented *Chlorine Content in and Life of Plastic Encapsulated Micro-Circuits* [8].

Failure mechanisms caused by chlorine contamination are mainly, aluminum corrosion and purple plague, and to some extent gold migration from the bond wires. However, during the last few years the quality of plastic encapsulated micro-circuits has been considerably improved.

Ritz, Stacy and Broadbent presented *the Microstructure of Ball Bond Corrosion Failures* in 1987 [9].

The microstructure of Au-Al ball bonds exposed to brominated molding epoxies under high temperature storage conditions has been examined. The main intermetallic phases found in the ball bonds after aging are Au_4Al and Au_5Al_2. At the periphery of the bond Au_2Al, AuAl and $AuAl_2$ phases can be found but in much smaller quantities. Au ball bonds aged in the presence of Br containing molding compounds undergo a microstructural and chemical change. The Au_4Al and Au_5Al phases develop a porous cellular structure consisting of Au cell walls and a cell interior containing Al_2O_3 (with traces of Br) plus precipitated crystallites of Au. Under appropriate stress conditions this corrosion and oxidation process produces enough Al_2O_3 to decouple the Au bond from the transformed intermetallic phases, causing bond lift failures.

In 1988, Muhib Khan, Homi Fatemi, Jeremias Romero and Eugene Delenia presented the *Effect of High Thermal Stability Mold Material on the Gold-Aluminum Bond Reliability in Epoxy Encapsulated VLSI Devices* [10].

"Use of thermally stable brominated CEN resins was found to significantly reduce wirebond failure. Presence of low molecular weight halogenated organic impurities and byproducts with labile C-Br or C-Cl bonds in the CEN resin had adverse effects on the Au/Al bond reliability. Halogenated organics were found to outweigh the beneficial effect brought about by superior thermal stability. Among the current generation of ion-free molding compounds, the interaction of organic brominated species with the Au/Al intermetallic is the predominant failure mechanism. The overall failure process was found to be controlled by the diffusion of small organic halides to the corrosion site."

"It is possible, by manipulating the chemistry of the brominated CEN resin, to make the molding compound more thermally stable. Through proper selection of the chemistry, process and material for brominating the CEN resin, it is possible to enhance the stability of the C-Br bonds in the resin network, minimize the formation of residual organic halides, and thus minimize the Au/Al failure."

In 1988, Syed Sajid Ahmad presented the *Impact of Residue on Al/Si Pads on Gold Bonding* [11].

"Gold wire bonds on Al/Si bonding pads were evaluated for bondability and bond reliability as a function of bonding pad over- and under-etch. Glass residues affect the bonds when the bonding pad is under-etched. When pads are overetched, residues deposited from the plasma affect the bonds."

"Bondability to pads with various layers of residue was evaluated by visual inspection of the bonds, by pull testing the bonds, and by measuring the electrical resistance of the bonds. Reliability was evaluated by monitoring electrical resistance of the bonds as a function of bake time."

"It is not surprising that bonds to under-etched pads do not stick and have low reliability. A small amount of over-etching was observed to improve the reliability of the bonds, but extended overetching decreased the lifetime slightly."

Yasuo Fukui, Koichiro Mukoyama and Hiromi Yamamoto were awarded Patent Number: 4,775,512, *Gold Line for Bonding Semiconductor Element* on October 4th, 1988 [12].

This invention relates to a wire bonding gold line used for connecting a semiconductor tip electrode with an outside lead portion. This gold wire characteristically contains gold (Au) having a purity level 99.996-99.9995 wt % and 0.0032-0.008 wt % by total weight of said gold wire of germanium (Ge) so that the electric resistance of the wire bonding gold line is reduced, while the mechanical strength such as tensile strength and high temperature strength is excellent and a gold ball can be shaped into the form of a nearly perfect sphere in the connection of the tip electrode.

David P. Szczublewski's (1989) thesis included a brief review of intermetallics [13].

"There are three well-known types of intermetallic compounds; valency, electron, and interstitial compounds .... Valency compounds are formed from metals that differ electrochemically. They obey the normal chemical rules of valency and are bonded together essentially by ionic or homopolar forces, or a combination of both. Electron compounds are formed from metals that are not markedly different electrochemically, but correspond to certain electron configurations. Interstitial compounds are formed from metals that have pronounced differences in atomic diameters. The smaller diameter atom to one metal is arranged between the spaces of the larger diameter atoms of the second metal..."

"The intermetallic compounds occur initially at the interface of the two metals and, as a result of the interdiffusion rates of the metals, can propagate within the metals themselves ... The propagation of the intermetallic compounds can be accelerated by large time-temperature products since high temperatures usually increase the interdiffusion rates of the metals…"

In 1989, Muhib M. Khan, Homi Fatemi, Jeremias Romero and Eugene Delenia presented *Enhancement of Gold-Aluminum Wirebond Reliability in Plastic Encapsulated Very Large Scale Integration (VLSI) Devices through C-Br Bond Stabilization* [14].

Cresolic epoxy novolac resins brominated by specially tailored brominating agents to impart a high C-Br bond energy remarkably enhanced thermal stability of the encapsulants and gold-aluminum wirebond reliability.

In 1989, Klaus-Jürgen Range showed that Au_5Al_2 actually was Au_8Al_3 [15]. It was indicated to be so in very early studies too, one such study being Heycock and Neville's (1900) *Gold-Aluminium Alloys* [16].

The intermetallic phase Al_3Au_8 was synthesized from the elements by high pressure high temperature reaction in a modified Belt-type apparatus. Synthesis experiments and structure determination show that the stoichiometry of the phase is indeed Al_3Au_8 and not Al_2Au_5 as previously suggested. There is no detectable range of homogeneity.

In 1989, Clatterbaugh and Charles Jr. presented *the Effect of High Temperature Intermetallic Growth on Ball Shear Induced Cratering* [17].

A mechanism that explains the high incidence of silicon cratering when thermosonic gold ball bonds are sheared from aluminum metallized pads over silicon and SiO_2 is proposed. The mechanism is based on the transmission of energy from the ball shear ram through the rigid intermetallic weld to the underlying dielectric. The concentration of this transmitted energy in the form of stress concentrators was estimated, using finite-element modeling, to be highly dependent on ball and weld geometry. This failure mode does not require that the substrate be cracked initially through the use of improper bonding conditions, as cratering was observed to occur even though the range of bonding parameters used in this study produced no visible damage to the underlying dielectric material. The above mechanism may be partly responsible for shear-induced cratering in plastic-encapsulated packages. Recommendations are made for reducing this effect by altering bonding parameters, pad metal thickness, and cure schedule.

In 1989, Wang, Fritz and Mendoza wrote in the *Chemistry of Stable Brominated Epoxies* [18].

> "In order to provide the required flame retardancy to the molding compound, an encapsulated formulation usually contains brominated resins and antimony oxide. The brominated resins used in the encapsulated formulation are mainly tetrabromobisphenol A (TBBPA) based epoxy resin or brominated epoxy novolac. These bromine-containing additives were reported to cause bond degradation at high temperature through accelerated void formation in the gold-aluminum intermetallic phases ([1, 6, 9, 19])."

> "In electronic encapsulation applications, epoxy derivatives of novolacs containing meta-bromo phenol have exhibited substantially better physical and performance properties compared to the conventional tetrabromobisphenol A epoxies and brominated epoxy novolacs. The meta-bromo phenol moiety contributes the expected improvements in thermal and hydrolytic stability to the formulation while providing fire retardancy properties ([21])."

In 1990, Corcoran, King, de Lanerolle and Kim studied grain boundary diffusion in comparison to mass or bulk diffusion, which showed non-$t^{1/2}$ dependence in the growth of a silicide layer [22].

In the same year, Maiocco, Smyers, Kadiyala and Baker (1990) published *Intermetallic and Void Formation in Gold Wirebonds to Aluminum Films* [23].

> Gold ball bonds attached to either pure Al films or Al films with Cu and Si additions were annealed at temperatures in the range [of] 77–277 °C for periods of up to 3,000 h[ours] and then shear tested. The resulting fracture surfaces were observed to change progressively with time and temperature of annealing. The intermetallic phases and voids present after each annealing treatment were characterized by microscopic examination of the bond cross sections. Shear test results are discussed in terms of the observed microstructures.

Maiocco, Smyers, Munroe, and Baker (1990) published a study on the *Correlation Between Electrical Resistance and Microstructures* in gold wire bonds on aluminum films [24]. Their data showed that the resistance change obeyed the $t^{1/2}$-law only for low resistance values up to 8 mΩ, and did not hold as soon as the temperature-time combination increased, giving rise to larger resistance values.

> "Electrical resistance measurements were performed on gold wire bonds to aluminum films, and the results were correlated to microstructural changes.
>
> 1) Resistance increases of less than 8 mΩ can be attributed to the growth of high resistance intermetallic phases, growth of a peripheral intermetallic ring around the bond, and depletion of the Al metallization around the peripheral ring.
> 2) The measured resistance increases, ΔR, analyzed as the difference from the mean of unannealed samples of the same metallization type, exhibit the following behavior up to 8 mΩ as a function of time (*s*) and temperature (*T*)
>
> $$\Delta R = C \cdot \exp(-E_a/kT) \cdot (t)^n$$

> ...where C = 10+/- 1.Ω/h^n , the activation energy E_a = 0.40±0.01 eV/atom, and n = 0.40±0.02. The confidence bounds given correspond to a 95 % confidence limit on the data."

In Blish and Parobek's (1983) paper, titled, *Wire Bond Integrity Test Chip*, they wrote the following. Our objective was a non-destructive indication of failure. We were not, as yet, sophisticated enough to test for Fickian behavior.

> "3) Large resistance increases of 20–70 mΩ observed at the highest annealing temperatures (223 °C and 277 °C) coincide with edge voids which grow and eventually separate the central bond region from the peripheral ring. The measured resistance was finite even after the aluminum conduction path was severed, indicating conduction through the underlying titanium film."

In 1990, Krzanowski published *a Transmission Electron Microscopy Study of Ultrasonic Wire Bonding* [25].

> Transmission electron microscopy (TEM) was used to examine ultrasonic bonding of aluminum wire[s] to various metal substrates. A special technique for preparing cross-sectional TEM specimens [is described,] involv[ing] [sample] embedding in epoxy, [followed by] dimpling and electropolishing. The materials systems examined [included,] Al wire bonded to Al foil, [as well as] Al wire bonded to Al foil [with evaporated] copper or gold [layers]. [TEM analysis of the bond interfaces revealed that in some regions,] surface oxides and contaminants were dispersed, allowing [direct contact between the bare metals]. [In contrast, adjacent areas of the interface were filled with debris.]. The debris could be oxides, surface contaminants, or sheared metal particles. It also appears that the debris is dispersed to low-lying regions between asperities. In studies of the Al/Au/Al and Al/Cu/Al interfaces, microchemical analysis showed no evidence for diffusion or melting along the interface. Therefore, it is concluded that ultrasonic wire bonding does not involve significant heating along the bonding interface. However, it is shown that dynamic annealing of [Al] wire[s] can occur during bonding. The [results of the] experiment [we]re compared with previously proposed theories on the mechanism of bonding in ultrasonic welding.

In 1990, Guy V. Clatterbaugh, Harry K. Charles Jr. published *The Effect of High-Temperature Intermetallic Growth on Ball Shear-Induced Cratering* [26].

> "The finite element analysis studies on the various ball and weld geometries show the following:
>
> 1) The effect of gold-aluminum diffusion induced strains within the weld region are negligible compared to external shear forces;
> 2) The smaller the weld region, the more likely the underlying silicon will crater when shear tested;
> 3) The taller the ball bond the more likely the underlying silicon will crater when shear testing;
> 4) The stress fields for an annular type of weld is similar to that for a circular type of the same radius.

These studies suggest that a flatter bond with a large weld area (or large annulus) will be less prone to produce silicon cratering when shear tested.

The results of the etching experiments indicated that the occurrence of cracks on bare silicon or SiO_2 on silicon due to improper bonding parameters was not likely for parameters falling within the bonding window used for this investigation. [This shows that damage to the substrate from poor bonding conditions is not required for ball shear cratering to happen.] [In] the case of the (BPSG) [borophosphosilicate glass] on SiO_2, damage of the thin oxide coating was prevalent for almost all bonding conditions. The best set of bonding conditions to resist cracking in BPSG films was found to be the combination of low power, short dwell, high stage temperature, and high force. This is consistent with the results reported by Koch [(1986) [5]] and his co-investigators … who studied the effects of bonding parameters for thermosonic gold ball bonding on aluminum pads over phosphosilicate glass (PSG).

The results from the metallization thickness experiment do indicate a significant reduction in the [frequency] of ball shear-induced cratering as the metallization thickness is increased. However, since the etching analysis of untreated wire bonded samples showed no initial substrate damage as mentioned above, a cushioning effect of the additional metallization was ruled out. A more plausible explanation would be that the additional metal would prevent the alloying of the gold ball and aluminum to the underlying silicon substrate during the thermosonic scrub. This would prevent a rigid link to the substrate available to transfer shear energy to the underlying silicon substrate.

Results from the experiment constructed to study the effect of bonding conditions on ball shear-induced cratering indicated that the greater the force and power parameters or [consistency], the stronger the bond, the less likely the substrate was to crater when ball shear testing. This is [essentially] stating that the larger the weld (flatter bond) the less likely that cratering will occur. Once again, this is consistent with the results from finite element modeling and previous data indicating that the manufacture of larger, more robust bonds are less likely to cause cratering...

Some conclusions can be drawn from the information above, concerning the effects of the gold-aluminum intermetallic on the cratering effect in silicon and thermally grown SiO_2 ... As stated above, the effect of strains introduced due to structural misfit of intermetallic phases is of second order compared to those introduced by the ball shear ram. Also, no damage was observed for either the underlying silicon or SiO_2 for the full range of bonding parameters used in this study. Therefore, it must be concluded that the cratering effect observed here is not a result of initial substrate damage. Several factors point to the rigid intermetallic bond between the ball and the substrate as the sufficient cause for ball shear induced cratering. These factors include:"

"1) The absence of cratering in samples which had not been thermally annealed (i.e., there is little intermetallic formation, and thus the aluminum metallization will yield before much energy can be transmitted to the underlying substrate);

2) The delay in the onset of cratering for thicker metallizations (a longer period of time would be required to penetrate the thicker metallization and alloy with the underlying silicon substrate);
3) The effect is significantly less for pads over SiO_2 (AuA1 intermetallic does not form as good a bond to SiO_2 as it will to bare silicon); and,
4) As the weld area increases with additional intermetallic formation, the estimated stress concentration factors become significantly reduced, thus explaining the reduction in cratering in strong bonds after prolonged thermal aging."

"In summarizing this cratering effect, a thicker bond pad metallization, a larger weld area (high ultrasonic energy and stage temperatures), and flatter bonds (high force) w[ould] produce bonds that are more resistant to any form of shear-induced cratering for bond pads on SO_2. Weak bonds are to be avoided since they can crater at much lower shear values. Preliminary information suggests that for bonding above multilevel oxides such as PSG, BPSG, and other low-strength CVD [chemical vapor deposition] deposited oxides, thicker pad metallization and the combination of low power, short dwell, hot stage temperature, and high force will be required to produce reasonably strong bonds and reduce damage to the underlying dielectric layers."

In 1990, Anthony A. Gallo presented the *Effect of Mold Compound Components on Moisture-Induced Degradation of Gold-Aluminum Bonds in Epoxy Encapsulated Devices* [27].

High moisture-aging of devices with Au-Al bonds, molded in epoxy can lead to ball-bond degradation in much the same way as had been observed previously for high temperature-dry aging. High moisture degradation appears to occur much faster at lower temperatures than dry degradation, and as a result, this stress test may be useful for screening molding compounds. Antimony trioxide appears to be the overriding factor in most molding compounds in causing high-moisture degradation. Its effect in a dry environment is presently being reexamined. This may clarify whether Sb_2O_3 is the "unknown formulation factor" alluded to in previous dry testing (6,10). Elimination of Sb_2O_3 and the use of ion scavenging additives with non-DGETA [Diglyceryl Ether of Tetrahydro Anhydrohexitol] types of flame-retardants could markedly improve the moisture performance of epoxy molding compounds. In a wider sense, the use of ion scavenging additives to break the cycle of galvanic action [corrosion], where more electropositive metals (e.g., Al) corrode [degrade] in contact with [less reactive ones], [i.e.,] those lower in the electromotive series (e.g., Au) may benefit diverse electrical applications.

In 1990, Mikkelson and Johnson reported on *Failure Mechanisms in GaAs ICs: E/D-Mode Refractory Metal Gate Failure Mechanisms* [28].

gold wires were ball bonded to Al/Si/Cu bonding pads. Devices were found to exhibit extensive intermetallic formation after burn-in and high temperature testing. To reduce this intermetallic problem components were assembled with aluminum wire and ultrasonically wedge bonded. Though the study did not uncover intermetallic problems between the aluminum wire and the gold plated package land bonds, there is the potential for its existence. Barrier layers such as chromium and TiPt have been found to retard [slow] intermetallic formation. [Quoted from Dylis 1991].

Servais and Brandenburg presented a closer look at wire bonding in 1991 [29].

"Various types of die bonding from eutectic bonds to epoxy bonds were found, but this did not appear to be a significant factor in bond quality."

"Metallurgical studies of the high-quality bonds showed an almost perfectly formed intermetallic present at the gold aluminum interface. Cross sections of the poorer quality bonds showed a spotty to non-existent intermetallic layer."

To make good bonds, process needs to be optimized for the target surface Al, Al/Si, Al/Cu and Al/Cu/Si, etc.

T/C -50° to + 150°C showed higher wire pull failure rates for molded samples when compared to unmolded bonds.

"… high temperature storage (1,000 hours at 150 °C) … Scanning electron microscope evaluation showed no detectable indications of degradation as a result of the exposure (there were no signs of excessive intermetallic growth or voiding). Cross sections of bonds made to the various metallization systems showed similar reactions at the interface. In all four metallization systems there was a significant growth of the intermetallic. Although other authors have reported voiding and a decrease in bond strength after extended periods of time at elevated temperature, our testing did not show excessive intermetallic growth (purple plague) or voiding after 1,000 hours of 150 °C storage … Mechanical testing of wire bonds showed no degradation in bond strength."

- "Significant Au/Al intermetallic growth was seen on all metallization systems as a result of 1,000 hours of 150 °C storage, but no voiding, loss of bond strength, or other degradation was noted.
- Bonds with a uniform intermetallic formation had a stronger bond shear strength than bonds with spotty or non-uniform intermetallic."

Editorial Note: With current thermosonic bonding technology, significant IMC formation is not expected after 1,000 hours at 150 °C.

In an unpublished study conducted by Intel, Ahmad, the compiler of this collection, measured bond resistance over a three-year period at 150 °C and observed no significant increase.

In 1991, Pretorius, Vredenberg and Saris published *Prediction of Phase Formation Sequence and Phase Stability in Binary Metal-Aluminum Thin-Film Systems Using the Effective Heat of Formation Rule* [30].

the effective heat of formation rule is used to predict first phase formation in metal-aluminum thin-film systems and to predict subsequent phase sequence for thin metal films on thick aluminum or thin aluminum on thick

> metal substrates. The effective concentration at the growth interface is taken to be that of the lowest temperature eutectic (liquidus) for the binary system. Although the effective heat of formation rule may predict that formation of a certain phase would lead to the largest free energy change, this phase does not necessarily form at the moving reaction interface if it has difficulty to nucleate. By excluding phases with a large number of atoms per unit cell and which thus have difficulty to nucleate, the effective heat of formation rule successfully predicts first phase aluminide formation for all 15 metal-aluminum binary systems for which experimental data could be found. It is also shown how the effective heat of formation rule can be used to predict formation and decomposition of aluminide phases in contact with each other or in contact with their component metals.

Editorial Note: Thermodynamics can predict the ultimate state, but kinetics may permit a metastable version (i.e., thermodynamics tells us what the final state should be, but mechanisms can allow a temporary, less stable state to exist).

In 1991, Murcko, Susko and Lauffer [31] discovered a linear resistance drift in aluminum to gold ultrasonic wire bonds at temperatures 75–125 °C, akin to systematic reduction in the radii of numerous conducting regions. A 15-mΩ delta resistance was selected as the fail point, since interfacial resistance could not be measured on the samples.

> 1) The activation energy for drifting was determined to be 0.73 eV/atom.
> 2) A threshold was found below 75 °C where resistance drifting ceased.
> 3) Baking prior to bonding significantly reduced the drift.
> 4) Thin gold can be plated at high current densities and still be reliable.
> 5) Thick gold requires careful selection of plating parameters.
> 6) The system becomes electrically stable when a full intermetallic develops down to the nickel.
> 7) An electrical monitor was developed which can detect poor gold plating in 3 h[ours] at 160 °C.

Editorial Note: This involves Al wire bonded to thick film Au, so one should expect different phases and different kinetics to that of the system, if it were Au wire to Al thin film. Plated thick films are notoriously vulnerable to copious impurities, metallic and organic, and surface composition may be different than bulk material.

In 1991, Eileen A. Sullivan published *Thermal Degradation of Epoxy Novolac-Phenol Formaldehyde Novolac Resin Systems* [32].

> When a combination of amine and phosphorous initiators is used, … the mechanism is accelerated. An interaction between the two types of initiators causing an increase in ionic bromide, provides one explanation of the increased failure rates… It is recognized that this is one possible explanation, and this mechanism probably accounts only partially for the acceleration of the ball bond failure. A number of other factors such as the presence of rubber modifiers and other additives may contribute.

In 1992, Mary G. Priore and John P. Farrell published a technology review of plastic microcircuit packages [33].

> "A traditional problem for plastic encapsulated ICs [integrated circuits] is the formation of intermetallic compounds at the interfaces of dissimilar metals. Plastic package[d] technology utilizes gold wire bonded to aluminum bonding pads throughout the industry. Interfaces of Al and Au in the wire bond will form intermetallic compounds when exposed to high temperatures for a prolonged period due to the different mutual interdiffusion rates of gold and aluminum. The action of this mechanism is further enhanced by thermal cycling and moisture…"
>
> "A host of catastrophic wire bond failures resulted from this mechanism. The failure mechanism became known as "Purple Plague" because of the presence of dark purple areas on the metallization surrounding the bond. This mechanism is associated with Kirkendall void formation at the bond/pad interface. This failure mechanism is increased by the presence of contaminants such as Bromine in flame retardant epoxies and residual chlorinated impurities."
>
> This [issue] has not been considered a major [concern in recent years, thanks] to such improvements as lower temperature assembly processes, [including] epoxy attachment of both [the] die and substrate. However, the only way to [fully] eliminate th[is] mechanism is to use a mono-metallic system, [an option not available] with plastic packages. [Nevertheless, Footner (1987)] [34] reports the re-appearance of this mechanism in the industry, with examples and lists the following factors as leading to its reoccurrence:
>
> * Tendency among manufacturers and users to disregard purple plague as a present day failure mechanism, forgetting that for purely thermodynamic reasons, intermetallics will always be formed when using the mixed metal system Al-Au...
> * A neglect of the forgotten role of interface contamination...
> * The use of epoxy attach arterials has led to the interesting suggestion that this is a source of contamination which enhances the intermetallic formation
> * The presence of water in many of the plastics used for encapsulation, allowing hydrogen from the water to combine with oxygen along the bond interface and hence negate the latter's role in reducing the diffusion of the Al..."

In 1992, Kang presented *Gold-to-Aluminum Bonding for TAB Applications* [35, 36].

> In thermocompression bonding used in tape automated bonding (TAB) technology, the gold-to-aluminum interface is one of the most difficult bonds that can be formed, because a thin aluminum oxide layer inhibits bonding at low temperatures. The usual bonding temperature is, therefore, expected to be higher than 500 °C. This results in extensive growth of intermetallics. In an effort to reduce the bonding temperature of a TAB application, a laser-processed gold-plated balltape was used for thermocompression bonding to aluminum. The effects of hardness, thickness, and morphology of the gold layer on thermocompression bonding have been investigated. As a result, an

improved gold layer surface morphology has been developed by controlling the plating rate. The new bumped tape structure was bonded at temperatures from 500 °C down to as low as 350 °C.

Editorial Note: TAB never made it to the "big time" as it was beaten by Controlled Collapse Chip Connection (C4). Blish's estimate of TAB's fatal flaw was that its tape was difficult to design and expensive relative to either wire or C4.

In 1992, Uno, Tatsumi and Ohno presented *Void Formation and Reliability in Gold Aluminum Bonding* [37].

> "… the annealing environments exert a great influence on the voiding as well as the intermetallic formation…
>
> 1. When the ball bonds are annealed in N_2 gas, the shear strength at first increases due to the intermetallic formation and in a later stage decreases due to the void formation. The deterioration in strength depends strictly on the degree of the voiding. For annealing in vacuum, the shear strength has a tendency to continue increasing.
> 2. The formation of the intermetallic compound layers and the voiding by annealing in the gaseous environments are much different from those for vacuum annealing ... The vacuum annealing promotes the metallic bonding of gold and aluminum in the interface in comparison with annealing in the gases. The bonds annealed in vacuum ensure high reliability even when they are subsequently exposed to elevated temperatures in the gaseous environments."

In 1993, Shirley and De Guzman presented *Moisture-Induced Gold Ball Bond Degradation of Polyimide-Passivated Devices in Plastic Packages* [38].

> A new mechanism of gold ball bond degradation in plastic packages has been demonstrated. Polyimide die topcoat and moisture stress accelerate bond degradation. The mechanism is further accelerated by mis-targeted bonds and "low" bonding parameters.

In 1994, Haji, Nakashima and Yoshinaga, published *Investigation on the Gold-Wire Bondability for the IC Device* [39].

> "In order to develop a microstructure observing technique for wire-bonded interfaces, which is important in the IC industry, *thin films* of wire-bonded gold-aluminum samples were prepared by ion milling and microtome techniques and observed by transmission electron microscopy (TEM). The effect of organic contamination for wire-bonding and the dependence of wire-bondability on the thickness of *plated gold film* and the plating method were also investigated by TEM and shear tests. The results obtained are as follows:
>
> 1) In the films prepared with a microtome no *amorphous phase* is observed, whereas in the films prepared by ion milling a large part of the intermetallic compound formed on the interface has become amorphous.

2) At the whole interface of the cleaned samples the intermetallic compound is observed, whereas at the interface of the contaminated samples an organic contamination layer is observed instead of the intermetallic compound. The cleaned samples are generally higher in shear strength than the contaminated samples.
3) Voids are produced by heating along the gold intermetallic compound interface. The voids deteriorate the bond strength.
4) Without heat treatment the bond strength does not depend on either the gold-film thickness or the plating method.
5) The nickel (from the substrate) concentration at the film surface increases with increasing heating period, and as a result the bond strength decreases.
6) The bond strength of samples bonded after heat treatment recovers when a surface layer of the gold film is removed before wire-bonding. From these results, a suitable technique for TEM observation of the wire-bonded interface and the effects of contamination on wire-bondability are discussed."

Editorial Note: Thin films are usually pure because they're made by evaporation or sputtering. Plated gold is usually thicker and can have both organic and inorganic impurities. More importantly, the presence of amorphous phases suggests a sectioning artifact (i.e., the amorphous areas might just be due to how the sample was cut).

In 1995, Koeninger, Uchida and Fromm published *Degradation of Gold-Aluminum Ball Bonds by Aging and Contamination,* in which they described the slow change in resistivity at low exposure times using the relation $\Delta R \sim \sqrt{t}$. They observed that the degradation mechanisms during annealing of the gold-aluminum ball bonds could be subdivided into three stages. They attributed the first stage to the growth of the compound layer obeying a $\sqrt{t}$-law. Pores in the intermetallic spread during the second stage. In the third stage, vertical growth of intermetallic compounds stops and takes place only in the horizontal direction leading to interface separation. Thermal expansion or the Kirkendall Effect by themselves cannot explain the large number of pores seen in the micrographs. They appear to be a result of the placement of the reaction fronts between the small amounts of aluminum reacting, preferentially with the large amount of gold near the edge of the gold ball [40].

In 1995, Tsuge, Mizuno, Uno and Tatsumi studied *Gold Diffusion and Intermetallic Formation in Au/Al_2O_3/Al Film* and reported [41].

Diffusion across the Al_2O_3 film in the Au/Al_2O_3/Al film system (the Al film with 1 μm thickness was vapor-deposited on a SiO_2/Si substrate and exposed to the atmosphere to form a natural oxide layer, then the Au film was deposited on it) at temperatures between 25 °C and 500 °C has been studied by using Auger Electron Spectroscopy, X-ray Photoelectron Spectroscopy and electrical resistivity measurement. The temperature where Al is detected on the surface of the Au/Al_2O_3/Al system is 100 °C higher than that on the surface of the regular Au/Al system without the Al_2O_3 film. The activation energies for the intermetallic layer growth of the Au/Al_2O_3/Al system and the Au/Al system are 110 and 72 kJ/mol, respectively. The Al_2O_3 film formed by the exposure in air (ca. 3.2 nm in thickness) acts as a barrier for diffusion in Au/Al. In addition, we observed the SEM image of cross section of the Au/Al_2O_3/Al system. The Au-Al intermetallic layer is formed in the Al layer in the initial stage of Au/Al_2O_3/Al diffusion by Au diffusion through the Al_2O_3 film into the Al layer. On the other hand, we studied the effect of annealing environment on the diffusion for the Au/Al_2O_3/Al system by using ^{18}O as tracer for SIMS analysis. The Au-Al intermetallic layer grows in an island formation for the Au/Al_2O_3/Al system and when annealed in air, the number of islands decreases. Because, during heat treatment in air, the Al_2O_3 film is formed continuously by supply of O_2 to the Al_2O_3 film through Au film.

Onuki, Satou, Murakami, Morita and Yatsuo (1995) reported *a New Low Temperature Diffusion Bonding Technology between Large-Area, High-Power Devices and Internal Mo Electrodes using Au-Al Films* [42].

To realize large-area, high-power devices, low temperature diffusion bonding between Al electrodes on both sides of the device and Au-plated Mo internal electrode foils has been investigated. Bonding was feasible below 573 °K due to the formation of Au-Al intermetallic compound. Substantial reduction of the mounting force while keeping contact uniform was also possible. Reliability of the bond type devices is predicted from metallurgical viewpoint.

In 1997, G. G. Harman published a comprehensive review of wire bonding in microelectronics: materials, processes, reliability and yield [43].

In 1997, Iida Kizaki, Fukuda and Mori published *the Development of Repairable Au-Al solid Phase diffusion Flip-Chip Bonding* [44, 45].

The authors have developed a new repairable chip-on-glass (COG) bonding technique for liquid crystal display (LCD) panels. Gold (Au) bumps on an LSI chip were bonded directly to aluminum (Al) electrodes on a glass substrate by formation of Al-Au intermetallic compounds in the diffusion layer. The developed repairable bonding technique consists a of two-level bonding process. First, the chip was bonded at 250 °C. Partial interconnection could be obtained at the local contact portions between the Au bump and the Al electrode. If the electrical connection failed, the bonded chip was removed. There was a distribution of the area formed Al-Au intermetallic compounds at local contact portions for 250 °C bonding. Some areas formed Al-Au intermetallic compounds of the Al electrode were sometimes removed with the chip removal, and an underlying metal layer was locally exposed at the remained surface. Then, a new chip was bonded on the same Al electrodes under the same conditions at 250 °C. After obtaining the electrical connection, the second bonding was done at 350 °C. An $AlAu_4$ intermetallic formation was obtained by this bonding in the diffusion layer. Reliability tests of second bonded samples were carried out and the contact resistance between the Au bumps and the Al electrodes was measured by the four-probe resistance measurement. In the case that the exposed area ratio of the underlying metal layer was less than 30 % of bonding area for each Al electrode, the stable electrical connection has been kept for a high temperature storage test and a thermal shock test. It was confirmed that a stable electrical connection had been obtained by the proposed repairable bonding process.

Thomas H. Ramsey and Cesar Alfaro published *High-Frequency Enhancement for Ambient Temperature Ball Bonding* in 1997[46].

new equipment enables ambient temperature ball bonding for the standard Au-Al and Au-Au processes, with shear values comparable to those found for 60 KHz bonds done at 200 °C or greater.

In 1997, Mori, Kizaki and Saito published *Au-Al Solid-Phase Diffusion Flip-Chip Bonding* [47].

A new flip chip bonding technique using Au-Al solid phase bonding was developed. In this technique, [a] layer of Au and Al is formed by connecting a driver IC with au bumps to an Al wiring board. In the conventional flip chip mounting method, the solder bumps formed on the LSI [large-scale integration] electrode are connected to the board by the reflow method. This method is difficult to apply to metal with weak solder wettability such as Al wiring. Furthermore, the reflow method is difficult to apply to electronic equipment constructed with material that has low heat resistance such as liquid crystal display (LCD) modules. Therefore, assuming the application to LCD panels with Al wiring, an investigation of connection conditions, analysis of heat transmission, and identification of the Au-Al intermetallic compounds were performed. The results show that connection is mechanically and electronically stable, the LCD panel is not affected by the connection conditions such as temperature, the intermetallic compounds are Au_5Al_2 and Au_4Al, and the final product is Au_4Al. Furthermore, an LCD module was fabricated by bonding a driver IC with 80 μm connection pitch to an LCD panel using Au-Al solid phase diffusion bonding.

In 1997, Fukudaet, Iida, Kizaki and Mori published *an Investigation of Bonding Mechanism for Au-Al Solid-Phase Diffusion Flip-Chip Bonding* [48].

Au bump and Al wire can be bonded by using the solid-phase diffusion between Au and Al. This Au-Al solid diffusion flip chip bonding technique was first used in connecting liquid display devices to the driving IC circuit and is becoming attractive as a method for interconnect between other types of semiconductor chips and substrates. The samples are evaluated and analyzed at different bonding temperatures, and the relationship between reliability and bonding interface is investigated. The results indicate that the reliability is dependent on the bonding temperature during storage of the samples at high temperature[s]. Since the bonding occurs when the solid diffusion reaction is almost in balanced condition in the case of high-temperature bonding, the bonding is reliable. When the bonding is formed before the solid diffusion reaction reaches balanced condition in the case of low-temperature bonding, if storing the sample at high temperature, the bonding deteriorates since the diffusion reaction persists at high temperature.

Hideo Nakajima published an overview of Earnest Kirkendall's work with respect to the Kirkendall Effect in 1997 [49].

The following year in 1998, Watanabe and Tabuse presented *Surface Contamination of Bonding Pads Incapable of Au Wire Bonding* [50].

> The spatial distribution of contamination in micro-areas of bonding pads in ICs and hybrid ICs is evaluated and graphically visualized by a micro-reflectometer of polarized light. Distributions of surface roughness, surface dielectric constant and the thickness of contaminant layers are calculated by non-classical polarized light reflectance theory. It is shown that the reflectance ratio R_s/R_p of s- and p-light is an excellent indicator of surface contamination, and its variation is proportional to weakening of the bond strength. It is concluded that this system can be used in microdevice production lines to monitor accidental or inevitable pollution of micro-components such as bonding pads because of its speed, cleanliness, compactness, and low cost.

Editorial Note: The s-light is polarized in the surface plane, while the p-light has a polarization partially into the surface plane (this technique is similar to ellipsometry).

In 1998, Imasato, Araki, Shimizu and Ohno presented the *Effect of Mold Resin on Reliability in Gold-Aluminum Bonding* [51].

1) Gold-aluminum wire bonds in LSI [were] drastically degraded by [the] corrosion of [the] Au-Al intermetallic compound.
2) When devices were annealed at [a temperature of] 200 °C for 100 h[ours] in [an] encapsulated condition, the corrosion product beg[an] to form at the interface, between [the] gold and Au-Al intermetallic compound.
3) In the case of the decapped condition, when the initial shear strength [wa]s high enough, this shear strength [remained] high after annealing at 200 °C for 100 h[ours]. But even this device was degraded drastically after annealing at 200 °C for 100 h[ours] in encapsulated condition.
4) The cause of degradation [wa]s [theoretically] due to [a] corrosion product formed between [the] gold and Au-Al intermetallic compound.
5) The activation energy of this degradation behavior was calculated [at] 110 KJ/mol (1.14 eV/atom), which was similar to that of [the] Au-Al intermetallic layer growth.
6) [The c]orroded intermetallic compound was Au_4Al, and [it formed a brittle, net-like structure].

Editorial Note: Corrosion usually denotes oxidation, whereas this "corrosion" is likely an intermetallic formation (i.e., whereas corrosion typically refers to oxidation, the 'corrosion' observed here is more likely due to intermetallic formation).

In 1998, Saito, Iida, Kizaki, Fukuda and Mori published *an Investigation of Stable Bonding for Au-Al Solid Phase Diffusion Bonding Techniques* [52].

> The authors have been developing a gold-aluminum (Au-Al) solid phase diffusion flip chip bonding technique. The Au bump on the IC was bonded directly to the Al electrode on the glass substrate by formation of an Au-Al

diffusion layer. This paper reports the reliability dependence on the Al layer thickness and analyses of the interfaces produced. The results confirm that it is possible to achieve initial electric contact. In the case of the bonding of a 350 nm Al layer, the intermetallic compound has the highest Au_4Al content. The bonding is stable after being subjected to a 125 °C atmosphere for 1,000 hours. The bonding of a 1,000 nm Al layer produces several kinds of Au-Al compounds in the intermetallic layer and the unreacted aluminum layer remains. This bonding condition does not yield sufficient reliability after being subjected to a 125 °C atmosphere for 1,000 hours. The difference in composition of the intermetallic layer affects the bonding reliability. When the Au_4Al content was at its highest level, a high reliability bond was obtained.

In 1999, Mori, Fukuda, Kizaki, Iida and Saito published *an Investigation of Stable Bonding for Au–Al Solid Phase Diffusion Bonding Techniques* [53].

By using a chip that has Au bumps and a substrate that has only a varying Al film thickness, initial bonding strength is made constant in Au-Al solid-phase diffusion bonding. On the other hand, by changing the Au-Al intermetallic compound formed during bonding, a relationship is obtained between the formed intermetallic compound and bonding reliability. Samples obtained when installing ICs under the same conditions on substrates with Al film thicknesses of 350 nm and 1,000 nm were left for 1,000 hours at 125 °C and the bonding strength and connection resistance were measured. Immediately after bonding, there was no meaningful difference between them. However, after 1,000 hours, reduced bonding strength and increased connection resistance were observed in the sample whose Al film thickness was 1,000 nm whereas a stable connection was obtained in the sample whose Al film thickness was 350 nm. The difference in reliability in a high-temperature environment results from differences in the Au-Al intermetallic compounds formed at the time of bonding and by the existence of unreacted Al. In the sample whose Al film thickness was 350 nm, there was no Al at the junction at the time of bonding, so that the final product was predominantly Au_4Al. On the other hand, in the sample whose Al film thickness was 1,000 nm, Al existed at the junction, so that several Au-Al intermetallic compounds were formed which subsequently degrade by diffusion reactions at high temperature.

In 1999, Suresh Kumar, Frank Wulff and Klaus Dittmer presented *Degradation of Small Ball Bonds due to Intermetallic Phase (IP) Growth* [54].

Kirkendall voids are an inherent part of intermetallic growth. Poorly bonded samples show [a] high concentration of large voids along the Au-IP interface, causing a drop in strength after thermal aging. However, well bonded samples exhibit very low void concentrations at the fully grown state resulting in strong, reliable bonds after long durations at elevated temperatures.

In 1999, Dittmer, Kumar and Wulff presented the *Influence of Bonding Conditions on Degradation of Small Ball Bonds due to Intermetallic Phase (IP) Growth.* Dittmer et al. (1999) also suggested that the diffusion of Au causes the transformation of some of the Au-Al phases to Au_4Al and Au_8Al_3 phases [55].

Uno, Terashima, Onoue and Tatsumi presented *Improvement in Thermal Reliability in Gold-Aluminum Bonds Encapsulated in Bi-phenyl Epoxy Resin* in 1998 [56].

> Au wire/Al pad bond degradation has become a major problem, because of the use of molding resin with low thermal stability (bi-phenyl epoxy resin) and the use of IC devices under high thermal environments. It is therefore important to secure the Au/Al bond reliability under high temperatures. The lifetime to bond failure for bi-phenyl epoxy molding became shorter than that for cresol novolac epoxy. The failures were caused by the corrosion reaction of Au-Al intermetallics with bromine (Br) contained in the resin compounds. It was clarified that the reactive intermetallic was the Au_4Al phase formed in the bond interface. The governing factors of the bond corrosion were investigated, such as the resin compound and gold wire material. In particular, impurities in the gold wire could affect the Au-Al intermetallic growth and therefore retard the corrosion. The use of the alloyed wire was effective in improving the bond reliability.

In 1999, Nicewarner studied *Interconnect Resistance Characteristics of Several Flip-Ship Bumping and Assembly Techniques*. Resistance measurements for Au-Al contact showed non-Fickian progress patterns [57].

In 1999, Uno and Tatsumi studied *Corrosion of Intermetallic Layer and Reliability of Au/Al Bonds* and reported that bond strength degradation was accompanied by significant resistance increase. They found that the affected layer was proportional to annealing time, indicating that the phenomenon was not controlled by diffusion [58].

Uno and Tatsumi published *Thermal Reliability of Gold-Aluminum Bonds Encapsulated in Bi-phenyl Epoxy Resin* in 2000 [59].

> The Au-Al bond reliability for BP resin was much lower than that for OCN [orthocreosol-novolac] resin. The activation energy of bond failure for BP resin was small compared to OCN resin.

Editorial Note: This information too qualitative to be useful. Also, corrosion is usually taken to imply water damage or oxidation, neither of which is present for Au-Al bond degradation. However, the authors correctly identified the terminal phase as Au-rich Au_4Al.

> The corrosive intermetallic was identified as Au_4Al phase and the corrosion products were Au precipitation and Al oxide.
>
> The impurities in Au wire would affect intermetallic growth and result in retarding [slowing] the corrosion. The use of the alloyed wire was effective in improving the bond reliability.

REFERENCES

[1] S. S. Ahmad, R. C. Blish II, T. J. Corbett, J. L. King and C. G. Shirley, "Effect of Bromine in Molding Compounds on Gold-Aluminum Bonds," *IEEE Transactions on Components, Hybrids, and Manufacturing Technology CHMT-9,* vol. 9, no. 4, pp. 379–385, 1986.

[2] R. A. Fouracre, "Electron Microscope Observations of Chemical Diffusion in the Al/Au System," *Thin Solid Films,* vol. 135, pp. 189–201, 1986.
[3] N. J. Noolu, N. M. Murdeshwar, K. J. Ely, J. C. Lippold and W. A. Baeslack III, "Partial Diffusion Reactions and the Associated Volume Changes in Thermally Exposed Au-Al Ball Bonds," *Metallurgic Materials Transaction A,* vol. 35A, p. 1273–1280, 2004a.
[4] N. Noolu, N. Murdeshwar, K. Ely, J. Lippold and W. Baeslack III, "Phase Transformations in Thermally Exposed Au-Al Ball Bonds," *Journal of Electronic Materials,* vol. 33, no. 4, p. 340–352, 2004b.
[5] T. Koch, W. Richling, J. Whitlock and D. Hall, "A Bond Failure Mechanism," in *IRPS,* pp. 55–60, 1986.
[6] M. M. Khan and H. Fatemi, "Gold-Aluminum Bond Failure Induced by Halogenated Additives in Epoxy Molding Compounds," in *Proceedings of the 1986 International Symposium on Microelectronics*, Georgia World Congress Center, Atlanta, Georgia, October 6–8, 1986 pp. 420–428.
[7] P. K. Footner, B. P. Richards, C. E. Stephens and C. T. Amos, "A Study of Gold Ball Bond Intermetallic Formation in PEDs using Infra-Red Microscopy," in *24th Annual Reliability Physics Symposium*, April 1986 pp. 102–108.
[8] K. Gustapsson and U. Lindborg, "Chlorine Content in and Life of Plastic Encapsulated Micro-Circuits," in *Proceedings, 37th (1987) ECC*, 1987.
[9] K. N. Ritz, W. T. Stacy and E. K. Broadbent, "The Microstructure of Ball Bond Corrosion Failures," in *IEEE/IRPS*, 1987 pp. 28–33.
[10] M. Khan, H. Fatemi, J. Romero and E. Delenia, "Effect of High Thermal Stability Mold Material on the Gold-Aluminum Bond Reliability in Epoxy Encapsulated VLSI Devices," 12–14 April 1988 pp. 40–49.
[11] S. S. Ahmad, "Impact of Residue on Al/Si Pads on Gold Bonding," in *IRPS,* pp. 534–538, 1988.
[12] Y. Fukui, K. Mukoyama and H. Yamamoto, "Gold Line for Bonding Semiconductor Element," Patent 4,775,512, 4 October 1988.
[13] D. P. Szczublewski, "The Redesign of a Multielectrode Semiconductor Array Intended for Implantation into the Brain of a Rhesus Monkey," December 1989.
[14] M. M. Khan, H. Fatemi, J. Romero and E. Delenia, "Enhancement of Gold-Aluminum Wirebond Reliability in Plastic Encapsulated Very Large Scale Integration (VLSI) Devices Through C-Br Bond Stabilization," in *Polymeric Materials for Electronics Packaging, ACS Symposium,* pp. 429–443, 1989.
[15] H. B. Klaus-Jürgen Range, "Hochdrucksynthese und Kristallstruktur von Al_3Au_8," *Journal of the Less Common Metals,* vol. 154, no. 2 (15 October 1989), pp. 251–260, 1989.
[16] C. T. Heycock and F. H. Neville, "Gold-Aluminium Alloys," *Philosophical Transactions of the Royal Society of London. Series A, Containing Papers of a Mathematical or Physical Character,* vol. 194, pp. 201–232, 1900.
[17] G. V. Clatterbaugh and H. K. Charles Jr, "The Effect of High Temperature Intermetallic Growth on Ball Shear Induced Cratering," in *1989 Proceedings. 39th Electronic Components Conference, 1989. Proceedings., 39th*, 22–24 May 1989.
[18] C. S. Wang, D. B. Fritz and A. Mendoza, "Chemistry of Stable Brominated Epoxies," in *Polymeric Materials for Electronic Packaging and Interconnection*, ACS Publications, pp. 397–404, 1989.
[19] R. J. Gale, "Epoxy Degradation Induced Au-Al Intermetallic Void Formation in Plastic Encapsulated MOS Memories," in *22nd Annual Proceedings International Reliability Physics Symposium,* 1984.

[20] R. M. Fisher, "Interfacial Failure Modes During Tests of Wire Bond Strength," in *Material Research Society, Session K7.1, Spring Meeting, April 24–29, 1989*, San Diego, CA., 1989.

[21] D. B. Fritz and C. S. Wang, "Abstr. No. 161," in *196th ACS National Meeting, Division of Polymeric Materials Science and Engineering*, 1988.

[22] Y. L. Corcoran, A. H. King, N. C. de Lanerolle, and B. Kim, "Grain Boundary Diffusion and Growth of Titanium Silicide Layers on Silicon," *Journal of Electronics Materials,* vol. 19, pp. 1177–1183, 1990.

[23] L. Maiocco, D. Smyers, S. Kadiyala and I. Baker, "Intermetallic and Void Formation in Gold Wirebonds to Aluminum Films," *Materials Characterization,* vol. 24, no. 4, pp. 293–309, 1990.

[24] L. Maiocco, D. Smyers, P. Munroe and I. Baker, "Correlation between Electrical Resistance and Microstructures," *IEEE Transaction on Components, Hybrids, and Manufacturing Technology,* vol. 13, pp. 592–595, 1990.

[25] J. E. Krzanowski, "A Transmission Electron Microscopy Study of Ultrasonic Wire Bonding," *IEEE Transactions on Components, Hybrids, and Manufacturing Technology,* vol. 13, no. 1 (Mar 1990), pp. 176–181, 1990.

[26] G. V. Clatterbaugh, J. Harry and H. K. Charles Jr, "The Effect of High-Temperature Intermetallic Growth on ball Shear-Induced Cratering," *IEEE Transactions on Components Hybrids, and Manufacturing Technology,* vol. 13, no. 1 (March 1990), pp. 167–175, 1990.

[27] A. A. Gallo, "Effect of Mold Compound Components on Moisture-Induced Degradation of Gold-Aluminum Bonds in Epoxy Encapsulated Devices," in *IRPS*, 1990.

[28] J. J. Mikkelson and D. Johnson, "Failure Mechanisms in GaAs ICs: E/D-Mode Refractory Metal Gate Failure Mechanisms," 1990.

[29] G. E. Servais and S. D. Brandenburg, "Wire Bonding – A Closer Look," in *The 17th International Symposium for Testing & Failure Analysis, 11–15 November 1991*, Los Angeles, California, USA, 1991.

[30] R. Pretorius, A. M. Vredenberg and F. W. Saris, "Prediction of Phase Formation Sequence and Phase Stability in Binary Metal-Aluminum Thin-Film Systems Using the Effective Heat Formation Rule," *Journal of Applied Physics,* vol. 70, pp. 363–364, 1991.

[31] R. M. Murcko, R. A. Susko and J. M. Lauffer, "Resistance Drift in Aluminum to Gold Ultrasonic Wire Bonds," *IEEE Transactions On Components, Hybrids, and Manufacturing Technology,* vol. 14, no. 4 (December 1991), pp. 843–847.

[32] E. A. Sullivan, "Thermal Degradation of Epoxy Novolac-Phenol Formaldehyde Novolac Resin Systems," *Journal of Applied Polymer Science,* vol. 42, no. 5 (April 1991), pp. 1815–1827, 1991.

[33] J. P. Farrell and M. G. Priore, "Plastic Microcircuit Packages: A Technology Review," Reliability Analysis Centre, Rome NY, March 1992.

[34] P. K. Footner, "Purple Plague: Eliminated or Just Forgotten?," *Quality and Reliability Engineering International,* vol. 3, pp. 177–184, 1987.

[35] S. K. Kang, "Gold-to-Aluminum Bonding for TAB Applications," in *Electronic Components and Technology Conference, 1992. Proceedings, 42nd*, 18–20 May 1992a.

[36] S. K. Kang, "Gold-to-Aluminum Bonding for TAB Applications," *Components, Hybrids, and Manufacturing Technology, IEEE Transactions on.,* vol. 15, no. 6 (December), pp. 998–1004, 1992b.

[37] T. Uno, K. Tatsumi and Y. Ohno, "Void Formation and Reliability in Gold Aluminum Bonding," in *Proceedings Joint ASME/JSME Advanced Electronics Package,* W. T. Chen and H. Abe, Eds., pp. 771–777, 1992.

[38] G. Shirley and M. S. De Guzman, "Moisture-Induced Gold Ball Bond Degradation of Polyimide-passivated Devices in Plastic Packages," in *Reliability Physics Symposium, 1993. 31st Annual Proceedings., International*, Atlanta GA, 23–25 March 1993.

[39] H. Haji, H. Nakashima and H. Yoshinaga, "Investigation on the Gold-Wire Bondability for the IC Device," *Engineering Sciences Reports, Kyushu University,* vol. 15, no. 4 (March), pp. 329–42, 1994.

[40] V. Koeninger, H. H. Uchida and E. Fromm, "Degradation of Gold-Aluminium Ball Bonds by Aging and Contamination," *IEEE Transactions on Components and Packaging Technologies Hybrids Manufacturing Technology*, vol. 18, no. 4, pp. 835–841, 1995.

[41] A. Tsuge, K. Mizuno, T. Uno and K. Tatsumi, "Gold Diffusion and Intermetallic Formation in Au/Al_2O_3/Al Film," *Journal Japan institute Metals,* vol. 59, no. 11, pp. 1095–1102, 1995.

[42] J. Onuki, M. Satou, S. Murakami, T. Morita and T. Yatsuo, "A New Low Temperature Diffusion Bonding Technology between Large-Area, High-Power Devices and Internal Mo Electrodes using Au-Al Films," in *Proceedings 8th International Symposium on Power Semiconductor Devices and ICs (ISPSD '96),* pp. 317–320, 1996 (20–23 May 1996).

[43] G. Harman, Wire Bonding in Microelectronics, McGraw Hill, 1997.

[44] A. Iida, Y. Kizaki, Y. Fukuda and M. Mori, "The Development of Repairable Au-Al Solid Phase Diffusion Flip-chip Bonding," in *Electronic Components and Technology Conference, 1997. Proceedings., 47th 18–21 May 1997,* pp. 101–107, 1997a.

[45] A. Iida, Y. Kizaki, Y. Fukuda and M. Mori, "The Development of Repairable Au-Al Solid Phase Diffusion Flip-Chip Bonding," *IEEE Transactions on Components, Packaging, and Manufacturing Technology, Part C,* vol. 20, no. 4 (October), pp. 343–34, 1997b.

[46] T. H. Ramsey and C. Alfaro, "High-Frequency Enhancement for Ambient Temperature Ball Bonding," *Semiconductor International,* vol. 20, no. 9 (August), pp. 93–96, 1997.

[47] M. Mori, Y. Kizaki and M. Saito, "Au-Al Solid-Phase Diffusion Flip-Chip Bonding," *Electronics and Communications in Japan (Part 2. Electronics),* vol. 80, no. 4, pp. 44–52, 1997.

[48] Y. Fukuda, A. Iida, Y. Kizaki and M. Mori, "An Investigation of Bonding Mechanism for Au-Al Solid-Phase Diffusion Flip-Chip Bonding," *Electronics and Communications in Japan (Part II: Electronics),* vol. 80, no. 10 (October), pp. 9–17, 1997.

[49] H. Nakajima, "The Discovery and Acceptance of the Kirkendall Effect: The Result of a Short Research Career," *JOM,* vol. 49, no. 6, pp. 15–19, 1997.

[50] M. Watanabe and K. Tabuse, "Surface Contamination of Bonding Pads Incapable of Au Wire Bonding," in *Proceedings of the IEEE/CPMT International Electronics Manufacturing Technology Symposium,* pp. 187–193, 1998.

[51] E. Imasato, M. Araki, I. Shimizu and Y. Ohno, "Effect of Mold Resin on Reliability in Gold-Aluminum Bonding," in *Proceedings of the IEEE/CPMT Electronics Packaging Technology Conferences*, pp. 338–344, 1998.

[52] M. Saito, A. Iida, Y. Kizaki, Y. Fukuda and M. Mori, "An Investigation of Stable Bonding for Au-Al Solid Phase Diffusion Bonding Techniques," *Transactions of the Institute of Electronics, Information and Communication Engineers C-II,* vol. J81C, no. 7 (July), pp. 628–636, 1998.

[53] M. Mori, Y. Fukuda, Y. Kizaki, A. Iida and M. Saito, "An Investigation of Stable Bonding for Au–Al Solid Phase Diffusion Bonding Techniques," *Electronics and Communications in Japan (Part II: Electronics),* vol. 82, no. 2 (February), pp. 11–20, 1999.

[54] S. Kumar, F. Wulff and K. Dittmer, "Degradation of Small Ball Bonds due to Intermetallic Phase (IP) Growth," 1999.

[55] K. Dittmer S. Kumar and F. Wulff, "Influence of Bonding Conditions on Degradation of Small Ball Bonds Due to Intermetallic Phase (IP) Growth," in *Proceedings 1999 International Conference on High Density Packaging and MCMs, (Reston, VA: SPIE, 1999),* pp. 403–408, 1999.

[56] T. Uno, S. Terashima, K. Onoue and K. Tatsumi, "Improvement in Thermal Reliability in Gold-aluminum Bonds Encapsulated in Bi-phenyl Epoxy Resin," *Proceedings of the SPIE – The International Society for Optical Engineering 1999, 1998 International Symposium on Microelectronics,* vol. 3582, pp. 591–596, 1998.

[57] E. Nicewarner, "Interconnect Resistance Characteristics of Several Flip-Chip Bumping and Assembly Techniques," *Microelectronic Reliability,* vol. 39, pp. 113–121, 1999.

[58] T. Uno and K. Tatsumi, "Corrosion of Intermetallic Layer and Reliability of Au/Al Bonds," *Journal of Japan Institute Metals,* vol. 63, no. 3, pp. 406–415, 1999.

[59] T. Uno and K. Tatsumi, "Thermal Reliability of Gold-aluminum Bonds Encapsulated in Bi-phenyl Epoxy Resin," *Microelectronics Reliability,* vol. 40, pp. 145–153, 2000.

11 2000: Electrostatic Discharge Under Pad and Probe Marks

Modeling and simulation, followed by experimental validation, are integral to the development of engineering processes. Advancements in manufacturing have enabled progress in computing, which have in turn supported the development of new materials and processes, ultimately making today's state-of-the-art devices possible. These devices have significantly enhanced the quality of life in various ways.

In 2000, Mercado, Radke, Ruston, Tran, Williams, Yong, Chen and Chen presented *Reliability of Multi-layer Aluminum Capped Copper Interconnect Structures* [1].

> Driven for further silicon (size) reduction, wireless applications utilize copper interconnection and increase metal layer count from three to five layers. More aggressive ESD structures placed under the bond pads offer a significant opportunity for additional die area and cost reduction. Capping copper bond pads with aluminum was selected as the primary approach for probing and wire bonding of copper devices. There is an integral relationship between probe damage on the bond pads and subsequent wire-bondability. As the pad geometry decreases, the ratio of the area of probe damage to the bond pad size becomes proportionally larger, thereby reducing the available aluminum necessary to form reliable gold-aluminum intermetallic coverage. This paper describes probe and assembly processes developed for a fine pitch three-metal layer copper interconnect device with ESD structures placed under the bond pads. The relationship between probe conditions and wire-bondability were examined. Ball shear, wire rip and corresponding failure modes were evaluated at various read points of thermal aging studies to evaluate the integrity of ball bonds to the metal stack. Reliability assessment was also performed. Based on the investigations studying the relationship between the pad structures, probe and wire bond quality, recommendations were derived to ensure high quality, stable and reliable bonds for fine pitch wire bonding on multi-layer copper interconnect devices.

Editorial Note: These findings are impressive, however, it is important to mention that they are qualitative in nature, which makes them inherently abstract, making it difficult to extract "design rules" or other useful information.

DOI: 10.1201/9781003619178-11

In 2001, Chang, Pon, Hsieh and Chen published *Intermetallic Growth of Wire-bond at 175 °C High Temperature Aging* [2].

> Two types of samples were applied in this wire-bond evaluation work for an Al-pad diffusion process. One was [a] finished product, [prepared] for cross-sectional examination of [the] wire-bond, [while the other lacked] compound encapsulat[ion and was] used for the plane view examination of [the] wire-bond interface. [These samples were aged in air at 175 °C for durations ranging from 0 to 1,008 hours]. It is found that the normal product which was ready on board existed ~1 μm Au_2Al and Au_5Al_2 phases. The Au_2Al phase was then transformed to Au_5Al_2 phase in a 4 ~ 72 h[ours] aging period. The Au_4Al phase formed along with the Au_5Al_2 phase in 72 h[ours] to 240 h[ours] aging period showing some porosity within the reacted phases. The total phase thickness increased to ~4 μm after 240 h[ours] aging. The Au_4Al phase became the major phase after 336 aging h[ours]. The reacted phase layer thickness increased to ~6 μm and reached the steady state. A titanium-rich thin layer was also induced within the reacted phase layer. Three ternary phases, [namely,] $AlAu_2Ti$, AlAuTi and Al_2Au_2Ti are found in this titanium-rich layer. In particular, Al_2Au_2Ti could be a new phase.

In 2002, Noolu, Klossner, Ely, Baeslack and Lippold presented *Elevated Temperature Failure Mechanisms in Au-Al Ball Bonds* [3].

1. Theoretical analysis indicates that, with the exception of Au_4Al and $AuAl_2$, the growth of the remaining Au-Al intermetallic phases results in volumetric shrinkage. Additionally, except for $AuAl_2$, the reverse transformation of the Au-Al phases is associated with minimal volume change.
2. Since phases identified during thermal exposure were Au_4Al, Au_8Al_3 and Au_2Al, the phase transformations occurring across the ball bond result in volumetric shrinkage.
3. This volumetric shrinkage induces shrinkage stresses across the bond. Combined with the presence of a void line, these stresses contribute to the degradation of the Au-Al ball bonds, particularly due to the growth of Au_8Al_3 and Au_2Al phases.
4. These shrinkage stresses facilitate the nucleation and growth of cavities the void line.
5. Crack propagation into the bond occurs through the linking (coalescence) of sufficiently grown cavities at the void line.

Editorial Note: The "cracks" are actually coalesced Kirkendall voids.

In 2002, Wang, Xu, Howlader, Itoh and Suga presented *Reliability and Microstructure of Au-Al and Au-Cu Direct Bonding Fabricated by the Surface Activated Bonding* [4].

> SAB (Surface Activated Bonding) method has been introduced as the most appropriate interconnection method for the next generation of electronic packaging because of room temperature process and other advantages. Thus, it is important to study the reliability of SAB interconnection in long term life test.

In this paper, SAB interconnection interface between Au and Al or Cu during high temperature thermal aging was investigated. The degradation of interface microstructure, and some properties for the interconnection during aging process were studied to investigate the failure mechanism of the interconnection. IMC (Intermetallic compounds) layer of Au-Al or Au-Cu were found formed during thermal aging, and that caused the failure mode of the interconnection changing in shear test. The results reveal that SAB method is highly reliable compared with other high temperature bonding methods such as wire bonding and soldering as there is no IMC layer in bonding process for SAB whereas others have.

In 2003, Wei and Daud published *Mechanical and Electrical Properties of Au-Al and Cu-Al Intermetallic Layer at Wire Bonding Interface* [5].

Comparatively, the Cu-Al intermetallics layer is about two times harder than Au-Al. Surface hardness of the Cu-Al intermetallics layer before 1.5 mm of indentation is 90 HT compared to zero for the Au-Al intermetallics layer. Both intermetallic systems grow linearly, but the Au-Al intermetallic is 20 times thicker than Cu-Al at 0 h[our] of aging test. [Its [Cu and Au] intermetallics layer growth rate is 18.0 and 0.07 nm/h, respectively.] Intermetallics thickness growth of 1.5 mm acquired an additional interfacial shear force of 160 gf to overcome the bonding, while the threshold value for the Au-Al and Cu-Al bonding system is 3.8 and 0.14 mm, respectively. Interfacial shear force was reduced after these thickness values. Electrical resistance for the Au-Al system increases rapidly after 600 h[ours] in the aging test, while for the Cu-Al system, it stays almost constant when compared to the Au-Al system. These results attributed that the Cu-Al bonding system is relatively more stable and more reliable than the Au-Al system.

In 2004, Bochenek, Bober, Hauffe, Lukaszewicz and Langer published *Thermal Degradation of Joined Thick Au and Al Elements* [6].

In investigated joints of thick Au and Al elements, the degradation mechanism is different than in the joints made of Al or Au wires on thin (no more than 1 μm thick) Au or Al (respectively) films. For the thick elements, the diffusion zones are significantly greater as complete consumption of any material does not take place. As a result of thermal treatment, the strongly defected regions occur deeply in the substrate and in the wire. In [the] Au substrate, the line of Kirkendall voids appears, which can be transformed into shrinkage crack of substrate material in the diffusion area with enhanced fragility. In [the] Al wire near the bond, material loss occurs mainly at the grains border and decreases the wire mechanical strength. These mechanisms explain the significant strength and resistance changes in the joints associated with the thermally activated processes, which diminish their reliability.

In 2004, Wulff and Breach published *New Observations on Intermetallic Compound Formation in Gold Ball Bonds: General Growth Patterns and Identification of Two Forms of* Au_4Al [7],

> "... a third layer of intermetallic grows between Au_4Al and Au_5Al_2. Measurements and observations made with EDX and optical microscopy lead to the conclusion that the new compound is a different form of Au_4Al ..."
>
> "... wire chemistry and bonding conditions are found to affect the thickness of the intermetallic compounds, which suggests that the resistance of ball bonds during molding and operation can change."

In 2004, Chang, Hsieh, Martens and Yang published *Wire-Bond Void Formation During High Temperature Aging* [8],

> Probe marks and Al pad contamination are the main causes of initial voids that block alloy diffusion and slow down intermetallic growth. Annular voids are caused by the ultrasonic squeeze effect of thermosonic wire bonding. These bonding gaps may become pathways for halide species that corrode and degrade wire bonds. Minute voids are formed during the Au_4Al phase. The two Au_4Al *phase textures* in these voids may be due to different Au_4Al phase formation reactions or be related to grain boundary effects on the surface layer of the Au ball.

In 2004, Wulff, Breach, Stephan and Dittmer presented *Characterisation of Intermetallic Growth in Copper and Gold Ball Bonds on Aluminum Metallization* [9].

> While the characterisation of intermetallic coverage and intermetallic phase (IP) growth in gold ball bonding on aluminum is quite well understood, there is relatively little literature concerning the morphology and growth of IP's between Cu balls bonded on aluminum pad metallisation. The difference between Cu-Al IP growth compared with the well known Au-Al IP's has been studied mainly of larger wire diameter (35–50 μm) in the early 1980's. Cu wire ball bonding has been established for many years mainly for high power devices at wire diameters ≥38 μm and fine wire for discrete device applications. However, there is now interest in fine pitch Cu wire ball bonding at smaller wire diameters of 25 μm and smaller for high pin count applications, driven mainly by cost reduction. Development and optimisation of robust copper wire bonding processes for such applications requires an assessment of intermetallic coverage and Cu-Al intermetallic growth after isothermal aging. This paper describes the problems associated with coverage determination, some characteristics of Cu-Al and Au-Al intermetallic compounds and characterises the difference in the IP growth between Au-Al and Cu-Al. The relative merits of gold and copper ball bonding are also briefly discussed.

Editorial Note: Figure 11.1 shows the IMC thickness for Au and Cu on Al film. Although the Cu intermetallic compound layer is significantly thinner, both Cu and Au exhibit unusually low activation energy values (0.38 eV for Cu and 0.16 eV for Au). The apparent "advantage" of Cu over Au is that, Au is highly temperature-dependent (sensitive to temperature) due to difference in their activation energies.

In 2004, Xu, Breach, Sritharan, Wulff and Mhaisalkar published *Oxidation of Bulk Au-Al Intermetallics* [11].

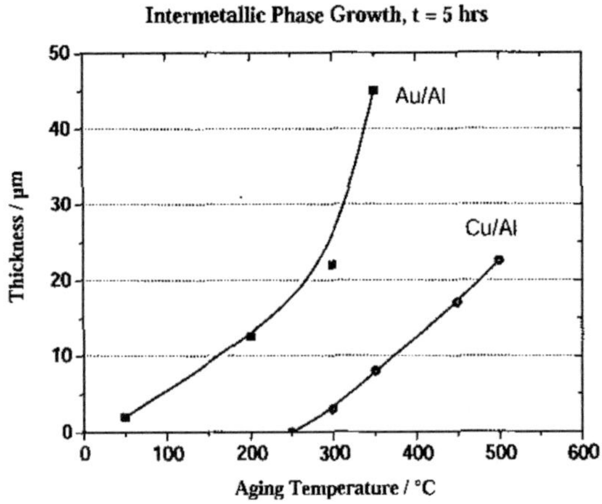

FIGURE 11.1 Phase growth of Au-Al and Cu-Al on diffusion couples after heat treatment [10].

Noolu, Murdeshwar, Ely, Lippold and Baeslack III wrote on *Partial Diffusion Reactions and the Associated Volume Changes in Thermally Exposed Au-Al Ball Bonds* in 2004 [12],

> A model to calculate the volume changes associated with phase transformations is presented. The molar partitioning of the reactants at each interphase boundary is taken as the basis for calculating volume changes across and lateral to the ball bond. Though the majority of the partial reactions describing the phase transformations show volumetric shrinkage, the actual volume change is dependent on the number of moles of each Au-Al phase that are actually undergoing the phase transformations. A criterion that governs the inward lateral growth of Au-Al phases was developed.

Noolu et ea. (2004) wrote the following in their study of *Phase Transformations in Thermally Exposed Au-Al Ball Bonds* [13],

> The as-bonded microstructure of a Au-Al ball bond essentially consisted of a reaction zone … between the Au bump and the bonded Al metallization. It is the growth of the reaction zone between the Au bump and the bonded Al metallization and also the non-bonded Al metallization during thermal exposure that gave rise to the various phase transformations. Au_4Al, Au_8Al_3, and Au_2Al are the predominant phases that grew across the ball bond until the bonded Al metallization is available to take part in the interdiffusion reactions. After the complete consumption of the bonded Al metallization, the Au-Al phases reverse transformed resulting in the formation of the Au_4Al phase in the entire reaction zone across the ball bond ... The lateral interdiffusion reactions resulted in the nucleation and the growth of all the Au-Al phases given by the phase diagram. ... It is possible for all the Au-Al phases to grow across a ball bond uninhibited as long as the bonded metallization is available. However, the supply limitation of the bonded metallization gives rise to reverse transformations where Al-rich

> phases transform to Au-rich phases and eventually result in the formation of the Au_4Al phase ... If infinite time were allowed, Au_4Al would dissolve; the extent of which is dependent on the solubility of Al in Au. No supply of Au lateral to the bond causes the reverse transformation of the Au_4Al phase, giving rise to the lateral growth of the remaining Au-Al phases. If infinite time were allowed, the lateral phase transformations would result in the formation of a phase that is dependent on the relative proportion of Au and Al present in the non-bonded metallization … and Au_4Al below the void line. Hence, the presence of a phase in a particular location of a ball bond is dependent on the time and temperature of thermal exposure."
>
> "The interdiffusion between Au and Al across a thermally exposed Au-Al ball bond causes the movement of the void line towards the Au bump and shows that Au interdiffuses faster than Al. Although the movement of the void line appears to be associated with movement of the Au_4Al/Au_8Al_3 interface, it is actually analogous to the Kirkendall marker movement."

Kim, Cho, Park and Lee's study of the *Effects of Pd Addition on Au Stud Bumps/ Al Pads Interfacial*, which covered reactions and bond reliability was published in 2004 [14].

> Early rapid growth of Au-Al IMCs during thermal aging treatment was slowed by the presence of a Pd-rich layer interrupting Au-Al interfacial reactions.

Editorial Note: Despite the insight, the users had to "take what we could get" in terms of what impurities were present in the wire.

In 2004, Kajihara published the *Analysis of Kinetics of Reactive Diffusion in a Hypothetical Binary System* [15],

> the thickness of the intermetallic compound increases in proportion to the square root of the annealing time. … the most predominant parameters controlling the growth rate are the diffusion coefficient and the solubility range of the intermetallic compound. The smaller the diffusion coefficient is, the slower the growth rate becomes. The effect of the solubility range on the growth rate is comparable to that of the diffusion coefficient on the growth rate. Thus, for the intermetallic compound with a narrow solubility range and a small diffusion coefficient, it will take a long time to reach to an observable thickness.

Li, Lei and Jue detailed the following in their 2004 report, *Studies of Microstructure Characteristics and Evolutions at the Bond Interface in Bonding Technology* [16],

> To enhance thermosonic bond development, an improved understanding is needed of the processes taking place at the bond interface. In this paper, lift-off and section characteristics at the interface of thermosonic bond were observed by using scanning electron microscope with EDS-test. It is obvious that the peeling underdeveloped bonds simulate a torus (or doughnut) with an unbonded central region and ridged peripheral region is bonded hardly. Bond strength is located between the severely ridged periphery and the non-adhering central area of the bond. For constant force and time, the ridged area of the

bond pattern increases when more power is applied. For constant force and power, the ridged location of the bonded region moves closer to the bond center with time. Au-Al and Au-Ag microstructures were identified by EDS-tests, and "Kirkendall" diffusibility in Au-Ag interface and inter-metallic compounds in Au-Al interface were confirmed. And these would be helpful for further research about thermosonic bonding.

Chad Robert Morgan published his work the *Investigation of Gold Aluminum Intermetallic Formations and their Degradation Mechanisms in Ball Grid Array (BGA) Multi Chip Modules (MCM)* in 2004 [17].

In many high reliability military and commercial avionics applications, the long-term reliability of the multi-chip modules used in the applications is of significant concern. Due to this concern, many high reliability applications continue to use hermetically sealed multi-chip modules to prevent exposure to the environment; however, since most high volume customers of electronic packages have turned to polymer-based encapsulants, these high reliability customers are finding it increasingly difficult to procure hermetically sealed packages. Therefore, the high reliability customers of hermetically sealed packages have been forced to evaluate the use of polymer-based encapsulant packages. One of the main concerns with the use of these non-hermetic packages is that they allow moisture to reach the surface of the integrated circuits of the package. In addition, the polymer-based encapsulants are historically known for containing halides that, when combined with the presence of moisture, may contribute to degradation of the wire bonds on the surface of the integrated circuits in the packages. This study attempted to evaluate the long-term reliability of packages incorporating polymer-based epoxy encapsulants that contained differing levels of halides, specifically chlorides. The study focused on the effects of the halides on the intermetallic phases formed during the wire bond process between the gold wire and aluminum bond pad. In order to evaluate these effects, multi-chip modules incorporating encapsulants with differing levels of chloride were subjected to up to 1,000 hours of Highly Accelerated Stress Testing (HAST) exposure that subjected the modules to elevated temperatures, elevated humidity levels, pressure, and electrical bias to evaluate the long-term reliability of the modules. The effects of this exposure and the focus of this study centered around corrosion of the intermetallic phases, intermetallic phase growth, Kirkendall voiding, and the effects of coefficient of thermal expansion (CTE) differences that lead to bond failure.

Okamoto published *Al-Au (Aluminum-Gold)* in 2005. Okamoto's work provided a brief review of the status of Al-Au phase diagrams [18].

Li et al. reported on *Atomic Diffusion Properties in Al/Ni and Au/Al Bond Interface* high density microsystem in 2005 [19].

Cross section in bonding point was produced by rubbing and polishing metallographic sample with metallographic inlaying powder. Cross section characteristics on ultrasonic bonded micro-interface (including Au/Al, Al/

> Ni) were tested by means of scanning electron microscope (JSM-6360LV) and EDS (EDAX2000). Lift-off characteristics at bonding micro-interface of thermosonic bond were observed by using SEM. Results show that atomic diffusion at the bonded interface was occurred each other. Thickness of Au/Al & Al/Ni interface was about 1–3 micron under ultrasonic and thermal energy. Inter-metallic compounds in Au/Al & Al/Ni interface were undiscovered under ultrasonic bonding in very short time that was the order of millisecond. Fracture morphology of lift-off interface was dimpled. Tensile fracture was generated by pull-test not only in bonded interface but also in basis material, and bonded strength at interface was strengthened by diffused atom from another side.

In 2005, Xu and Suga, reported on *Surface Activated Bonding – High Density Packaging Solution for Advanced Microelectronic System* [20].

> Surface activated bonding (SAB) was a recently developed high density interconnection method for advanced microelectronic system. SAB bonding was proposed based on the simple principle that two surface activated materials could be easily bonded together at the room temperature or low temperature. The surfaces were *activated by fast atom beam or plasma irradiation source* which has been proved helpful to remove the oxide and organic layer covering on the material surface. Chip-on-chip (COC), as an important application of SAB flip chip method, is discussed in this paper. To investigate the SAB bonding feasibility, some electrically plated Au microbumps which have a size of $40 \times 40 \times 30\ \mu m^3$, were prepared on the silicon chip and bonded with Au, Al or Cu pads with a thickness of about 0.3 μm on another silicon chip. Electrical measurement, shear test and thermal aging (423 °K) data demonstrated the fine performance for SAB interconnection. It was found even the bonded chip-chip system was die separated, SAB achieved Au bump-metal pad interconnection was remained well. As bonding is carried out at the room temperature or low temperature and the material is surface activated before, no intermediate layer or impurity particle will exist at the as-bonded interface. That, for the SAB bonded Au-Al sample, a homogeneous intermetallic compound layer was finally grown and formed at the Au-Al interface (Au-Al) without any voids after 500 hours thermal aging at 423 °K in air, which was completely different from the performance of Au-Al interface constructed by other interconnection techniques such as wire bonding or soldering flip chip method. The application of SAB flip chip bonding on chip-on-film and chip-on-flexible board (COF) was also investigated. SAB flip chip method realized a high bonding accuracy of about 0.5 μm for the above systems.

Editorial Note: Rich is even more skeptical of surface activated bonding (SAB) after learning that the surface activation relies on plasma or high energy ion bombardment. Except under Ultra High Vacuum (UHV) conditions, surface chemistry can change substantially in a very short time. Moreover, this study was conducted using Au microbumps (likely plated), with a level of purity that is probably quite different from that of spherical Au balls created using a hydrogen torch.

REFERENCES

[1] L. Mercado, R. Radke, M. Ruston, T. Tran, B. Williams, L. Yong, A. Chen and S. Chen, "Reliability of Multi-layer Aluminum Capped Copper Interconnect Structures," in *Electronics Manufacturing Technology Symposium, 2000. Twenty-Sixth IEEE/CPMT International*, pp. 84–93, 2000.

[2] H. S. Chang, J. X. Pon, K. C. Hsieh and C. C. Chen, "Intermetallic Growth of Wire-bond at 175 °C High Temperature Aging," *Journal Electronic Materials,* vol. 30, no. 9, pp. 1171–1177, 2001.

[3] N. Noolu, M. Klossner, K. Ely, W. Baeslack and J. Lippold, "Elevated Temperature Failure Mechanisms in Au-Al Ball Bonds," in *International Symposium on Microelectronics,* pp. 478–483, 2002.

[4] Q. Wang, Z. Xu, M. Howlader, T. Itoh and T. Suga, "Reliability and Microstructure of Au-Al and Au-Cu Direct Bonding Fabricated by the Surface Activated Bonding," in *Electronic Components and Technology Conference, 2002. Proceedings. 52nd*, 2002.

[5] T. C. Wei and A. R. Daud, "Mechanical and Electrical Properties of Au-Al and Cu-Al Intermetallic Layer at Wire Bonding Interface," *Journal of Electronic Packaging,* vol. 125, no. December, pp. 617–620, 2003.

[6] A. Bochenek, B. Bober, W. Hauffe, M. Lukaszewicz and E. Langer, "Thermal Degradation of Joined Thick Au and Al Elements," *Microelectronics International,* vol. 21, no. 1, pp. 31–34, 2004.

[7] C. D. Wulff and F. W. Breach, "New Observations on Intermetallic Compound Formation in Gold Ball Bonds: General Growth Patterns and Identification of Two Forms of Au_4Al," *Microelectronics Reliability,* vol. 44, pp. 973–981, 2004.

[8] H.-S. Chang, K.-C. Hsieh, T. Martens and A. Yang, "Wire-Bond Void Formation During High Temperature Aging," *IEEE Transactions on Components and Packaging Technologies,* vol. 27, no. 1 (March), pp. 155–160, 2004.

[9] F. W. Wulff, C. D. Breach, D. Stephan and K. Dittmer, "Characterization of Intermetallic Growth in Copper and Gold Ball Bond on Aluminum Metallization," in *Electronic Packaging Conference,* pp. 348–353, 2004.

[10] A. Bischoff and F. Aldinger, in Proceedings of the 34th Electronic Components Conference, May 14–16, New Orleans, Louisiana, 1984.

[11] C. Xu, C. D. Breach, T. Sritharan, F. W. Wulff and S. G. Mhaisalkar, "Oxidation of Bulk Au-Al Intermetallics," *Thin Solid Films,* vols. 462–463, pp. 357–362, 2004.

[12] N. J. Noolu, N. M. Murdeshwar, K. J. Ely, J. C. Lippold and W. A. Baeslack III, "Partial Diffusion Reactions and the Associated Volume Changes in Thermally Exposed Au-Al Ball Bonds," *Metallurgic Materials Transactions. A,* vol. 35A, pp. 1273–1280, 2004.

[13] N. Noolu, N. Murdeshwar, K. Ely, J. Lippold and W. Baeslack III, "Phase Transformations in Thermally Exposed Au-Al Ball Bonds," *Journal of Electronic Materials,* vol. 33, no. 4, p. 340–352, 2004.

[14] H.-J. Kim, J.-S. Cho, Y.-J. Park and J. Lee, "Effects of Pd Addition on Au Stud Bumps/ Al Pads Interfacial," *Journal of Electronic Materials,* vol. 33, no. 10, pp. 1210–1218, 2004.

[15] M. Kajihara, "Analysis of Kinetics of Reactive Diffusion in a Hypothetical Binary System," *Acta Materials,* vol. 52, pp. 1193–1200, 2004.

[16] J-H. Li, H. Lei and Z. Jue, "Studies of Microstructure Characteristics and Evolutions at the Bond Interface in Bonding Technology," in *High Density Microsystem Design and Packaging and Component Failure Analysis, 2004. HDP '04. Proceeding of the Sixth IEEE CPMT Conference on. 30 June–3 July 2004.*

[17] C. R. Morgan, "Investigation of Gold Aluminum Intermetallic Formations and their Degradation Mechanisms in Ball Grid Array (BGA) Multi Chip Modules (MCM) Thesis (M.S.)," Iowa State University, 2004.
[18] H. Okamoto, "Al-Au (Aluminum-Gold)," *Journal of Phase Equilibria and Diffusion,* vol. 26, no. 4, pp. 391–393, 2005.
[19] L. Jun-hui, H. Lei and Z. Jue, "Atomic Diffusion Properties in Al/Ni and Au/Al Bond Interface," in *High Density Microsystem Design and Packaging and Component Failure Analysis, 2005 Conference on.*, 27–29 June 2005.
[20] Z. Xu and T. Suga, "Surface Activated Bonding – High Density Packaging Solution for Advanced Microelectronic System," in *Electronic Packaging Technology, 2005 6th International Conference on*, 30 August–2 September 2005.

12 2006: Impurities and the Reliability of Gold Wire

The gold wire was bound to receive its due attention eventually, given that the gold-aluminum interface had long been the primary focus.

In 2006, Gam, Kim, Cho, Park, Moon and Paik reported on the *Effects of Cu and Pd Addition on Au Bonding Wire/Al Pad Interfacial Reactions and Bond Reliability* [1].

According to observations of the AuAl interfacial reactions, Cu and Pd additions to Au bonding wire formed a Cu-rich layer and a Pd-rich layer at Au and AuAl IMC interfaces after thermal aging. In Au-1 wt.% Cu wire, the slow growth of the Au_4Al phase, along with a Cu-rich layer that hindered Au diffusion, significantly slowed the AuAl interfacial reaction and IMC growth, ultimately reducing crack propagation. On the other hand, in Au-1 wt.% Pd wire, the AuAl interfacial reaction rate decreased due to a Pd-rich layer that inhibited Au diffusion, and accordingly, resulting in slower crack formation. The addition of Cu was the most effective in slowing crack propagation. In addition, STEM observations revealed that the Cu-rich layer was free of voids, whereas the Pd-rich layer with numerous voids in its lower half, served as a pathway for crack propagation.

Bond reliability testing proved that adding Cu and Pd improved bond reliability, and that, Cu was more effective for improving bond reliability than Pd was. The results of this bond reliability test corresponded well with the observations of AuAl interfacial reactions. Interfacial morphology and wire pull test results also indicated that cracks formed as a result of AuAl interfacial reaction were an important factor in determining bond reliability.

It was found that adding either Cu or Pd to the Au wire enhances bonding to Al.

Blish, Li, Kinoshita, Morgan and Myers 2006, said in 2006 [2].

> Au-Al InterMetallic Compounds (IMC) grow laterally (Al-rich phases) in a Fickian fashion with an activation energy of 1.0 eV, but vertical IMC thickness (Au-rich phases) grows functionally as a power law on time with a sub-Fickian exponent of ~¼, substantially smaller than what would be expected for bulk, lattice diffusion (½). We conclude from the IMC thickness time exponent that [the] Au-rich IMC growth process is limited by grain boundary diffusion. The best bond lifetime was seen for an intermediate thickness Al film. The activation energy and lifetime [of] Au-rich phase growth are each a strong function of wire impurity concentrations. We find that bond lifetime varies roughly as the

DOI: 10.1201/9781003619178-12

> square root of Pd, Cu, Pt and As concentrations, but lifetime is not a function of Be, Ca, Fe or Cr concentration. We find a mixture of Au_4Al and Au on the Al-rich side of the failing interface.

Editorial Note: This abstract should also have said that the most powerful additive to improve Au-Al bond lifetime is Arsenic (few ppm sufficient for 100X in "automotive" use), although Cu + Pt at large concentrations (thousands of ppm produce 5X improvement in "automotive" usage) were also effective.

In a rare moment of pride, I would like to place our own paper first after the marker for impurities in the wire. I hadn't realized until now how many other papers on same subject were published later in 2006. Dr. Blish noted that, it was humbling to see what we were ignorant of at the time.

In 2007 Wulff, Tok and Breach reported their results on a study of *Oxidation of* Au_4Al *in Gold Ball Bonds* [3].

In 2006, Klengel, Heideallee Knoll, Petzold, Wohnig and Schrapler studied *Intermetallic Compound Formation in Au/Al Thermosonic Wire Bonding during High Temperature Annealing at 150 °C as a Function of Wire Material* [4].

> the investigated Pd-doped 2N wire exhibited both the lowest IMC growth as well as defect formation rate in comparison to the additionally investigated 4N and 4.5N wires.

Editorial Note: The 2N means two nines pure, i.e., 99 %, while 4N would be 99.99 % pure (100 ppm total impurity).

In 2006, Blish, Li, Kinoshita, Morgan and Myers reported on *Gold-Aluminum Intermetallic Formation Kinetics* [2].

> "Au-Al intermetallic compounds (IMC) grow laterally (Al-rich phases) in a Fickian fashion with an activation energy of 1.0 eV, but vertical IMC thickness (Au-rich phases) grows functionally as a power law on time with a sub-Fickian exponent of ~¼, substantially smaller than what would be expected for bulk, lattice diffusion (½). We conclude from the IMC thickness time exponent that Au-rich IMC growth process is limited by grain boundary diffusion. The best bond lifetime was seen for an intermediate thickness Al film. The activation energy and lifetime for Au-rich phase growth are each a strong function of wire impurity concentrations. We find that bond lifetime varies roughly as the square root of Pd, Cu, Pt and As concentrations, but lifetime is not a function of Be, Ca, Fe or Cr concentration. We find a mixture of Au_4Al and Au on the Al-rich side of the failing interface."
>
> "While on the journey to reduce lateral growth of Al-rich phases, we noticed that the Au-rich phases were forming beneath the bond more rapidly than we expected."
>
> "… we found … that the "signal" for reliability as a function of Al thickness was weak at best."
>
> "One of the unexpected results was that the IMC growth kinetics and mechanism (the latter inferred from the time exponent) changed with Al thickness. Really thick Al films (3,000 nm) showed a IMC growth time exponent of ½,

characteristic of bulk, lattice diffusion, but we saw much smaller time exponents (~¼) for thin films (470–1,000 nm)."

"We found a statistically significant correlation between the ppm concentrations for various impurities added to the Au wires … we find that bond lifetime varies (empirically) as a power law on fcc impurity concentration."

"… the best wire had "modest" fcc-type impurity concentrations, yet the performance was superlative … the … unexpected element … responsible for the stellar performance of wire … turned out to be Arsenic! Arsenic's impurity exponent was also near ½. A plausible explanation for why As might be responsible is drawn from traditional practice to harden Pb, which has the same crystal structure as Au. That one would use Sb (Group V) to harden Pb, is analogous to using As (also Group V) to harden Au. We find a mixture of Au_4Al and Au near the failing interface."

Editorial Note: This abstract should also have said that the most powerful additive to improve AuAl bond lifetime is Arsenic (few ppm sufficient for 100X in "automotive" use), although Cu + Pt at large concentrations (thousands of ppm produce 5X improvement in "automotive" usage) were also effective.

In 2006, Hung, Lin, Chen, Wang and Hsiao reported on *the Characterization of Intermetallic Growth in Copper and Gold Ball Bonded on Thicker Aluminum* [5].

> 2N gold alloy wire (Purity: 99 %) is recommended to replace 4N gold wire (Purity: 99.99%) for improving 1st bond failure. Due to palladium-rich layer formation at the gold ball and intermetallic compound, the Pd barrier layer restrains the gold atom, which cannot continue to move to the Al metallization side to avoid crack propagation. However, gold wire with Pd element cannot overcome the crack propagation or void formation. Slower intermetallic compound property of Cu wire is investigated with thicker Al metallization. This paper describes the characteristics of Cu-Al and Au-Al intermetallic compounds and characterizes the difference in the intermetallic compound growth between Au-Al and Cu-Al for thicker Al metallization (~5 μm).

In 2006, Muller, Schrapler, Altmann, Knoll and Petzold reported on the *Influence of Intermetallic Phases on Reliability in Thermosonic Au-Al Wire Bonding* [6],

> Applying focused ion beam (FIB) techniques as well as scanning and transmission electron microscopy (SEM, TEM) alternative failure modes in addition to Kirkendall voids were found that could be divided into five morphological categories. These failures include [1] the missing of any Au-Al intermetallic compound (IMC) formation at the pads after wire bonding, [2] no uniform horizontal/vertical IMC formation after wire bonding, [3] micro cracks between the interface IMC and [4] Au ball during reliability stressing, micro cracks between the interface of the two IMCs during temperature aging and [5] finally corrosion of the interface IMC/Au ball during temperature aging or humidity testing. Although there is strong indication that part of these failure modes are related to front end process steps, the practical limitations resulting from separation between semiconductor manufacturing and assembly line require a back end solution. It is shown that the use of an improved high security (HS) or high

reliability wire (HR) contributes significantly to reliability improvement. The *wires contain special dopants* which reduce the IMC and defect growth. The influence of these wires on the reliability of bonds are explained and shown in selected examples.

Li, Li and Ji studied *Metallurgical Behavior of Au/Al Bond Interface during High Temperature Storage of Ultrasonic Wedge Bonding* in 2006 [7],

there was evident diffusion of Au and Al, and Au_5Al_2 intermetallic compound (IMC) formed first after aging for 10 days at 200 °C in atmosphere, and there was also some Au_2Al formed at the edge of the interface. After aging for 20 days, the IMC developed slightly, and also Kirkendall voids generated at the interface. After aging for 30 days, the IMC began to transform to other Au-Al IMCs, and it became Au_2Al IMC finally after aging for 40 days.

In 2006, Xu and Suga reported on *Room/Low Temperature Interconnection Technique on Micro-bump/Film for COC and COF System* [8],

In this paper, surface activated bonding (SAB), as a new high density bonding method and its application on the interconnection between metal micro-bump and metal film for the chip-on-chip and chip-on-flex/chip-on-film system, was introduced. Bonding feasibility of Au bump and Al film in the vacuum at room temperature, Au bump and Cu film, Au bump and Au film, and Cu bump and Cu film in air or nitrogen ambience with atmospheric pressure at low temperature were well demonstrated. Bonding shear strength of more than 15 MPa and electrical resistance below 7 mΩ proved a good performance of SAB process on the interconnection for electronic system. Thermal electrical and mechanical performance appearance of SAB interconnection illustrated an acceptable anti-heat ability of SAB bond even without the protection of under-fill layer. The obviously different bonding feasibility, reliability, and interface structure of Au-Al bond by SAB and wire bonding (WB) selected the advantages of SAB over the traditional bonding technique. SAB process could establish a dense surface contact between the surfaces of two bonding materials, which inhibited the potentially inhomogeneous formation and growth of intermetallic compound layer, thus help establish a stable mechanical and electrical performance for interconnection in the electronic system even after a long-time application (Zhonghua Xu et al., 2005). SAB was a new process for the interconnection in the microelectronic system. Compared with thermal compress, soldering flip chip process, etc., as of low/room temperature process characteristics, SAB was especially suitable for high density interconnection. Micro-bump/film for COC and COF system based on SAB technique were discussed. New generation SAB bonders were developed successfully. Bonding accuracy of about 1μm at room temperature in vacuum or low temperature in air or nitrogen ambience could be easily succeeded for the interconnection in packaging system. Optoelectronics, medical systems, or other high precise microelectronic systems were well developed based on the system of SAB technology.

In 2006, Alexander Teverovsky (the king of opaque acronyms) reported on the *Effect of High Temperature Storage in Vacuum, Air, and Humid Conditions on Degradation of Gold-Aluminum Wire Bonds in PEMs* [9],

- Considering encapsulating molding compound (MC) as an environmental factor, the results of this work show[s] that all environments, including oxygen, moisture, and MC, substantially decrease time to catastrophic Au/Al WB failure compared to vacuum conditions.
- The presence of MC during HTS decreases the characteristic life of WBs [wire bonds] by a factor of ~10 in vacuum (at 198 °C) and by a factor of ~2 in air (198 °C). In humid environments (150°C/85%RH) this decrease was estimated as a factor of ~3 to 5 for OCN [cresol novolac] and ~20 for BP MCs [biphenyl molding compounds]. This means that WBs in PEMs are especially vulnerable to possible manufacturing defects, thus requiring additional attention when commercial PEMs are to be qualified for high-reliability applications.
- HTS [high temperature storage, also known as baking] of PEMs in vacuum increases the characteristic life of WBs compared to air conditions on 5 % to 35 % only. However, for decapsulated devices, these variations are much more significant and at 198 °C time to the first WB failure increases in vacuum to more than 2,500 h[ours] compared to ~250 h[ours] in air. This effect is due to the thermal decomposition of MC, which provides a sufficient amount of oxidative agents to oxidize Au-Al intermetallics in PEMs and cause failures even in vacuum conditions.
- Characteristic times of WB failures in PEMs stored in air at temperatures from 190 °C to 200 °C vary from 500 to 1,000 h[ours] and their activation energies are 0.71 to 0.79 eV for OCN MC and 0.91 eV for parts in BP MC.
- WBs failed in decapsulated devices after HTS in vacuum separated along the intermetallic/barrier metal interface, whereas in all other environments the separation occurred along the intermetallics exhibiting a grainy structure of a gold-rich composition with a grain size of 100 to 400 nm.

In 2006, Jang, Duh, Takahashi and Su studied *Solid-State Reaction in an Au Wire Connection with an Al-Cu Pad during Aging* [10].

> In the As-assembled wire bond, $(Al,Cu)Au_4$, $(Al,Cu)_3Au_8$, and $(Al,Cu)Au_2$ IMCs were detected between Au wire and Al-Cu pad. After 168 h[ours] of aging at 150 °C, $Al_3(Au,Cu)_8$ formed at the interface of $(Al,Cu)_3Au_8$, and $(Al,Cu)Au_2$ in the center of the wire bond. Due to supply limitations of the thin AlCu pad in the center of wire bonds, $(Al,Cu)Au_2$ was completely transformed into $Al_3(Au,Cu)_8$ after depletion of the Al-Cu pad. Because $Al_3(Au,Cu)_8$ was in contact with chip, $Al_3(Au,Cu)_8$ was immediately transformed into $(Al,Cu)_3Au_8$. Therefore, the growth rate of $(Al,Cu)_3Au_8$ was most rapid before complete consumption of $Al_3(Au,Cu)_8$. After 500 h[ours] of aging at 150 °C, the Al-Cu pad, $(Al,Cu)Au_2$ and $Al_3(Au,Cu)_8$ were exhausted, and $(Al,Cu)_3Au_8$ was thus transformed into $(Al,Cu)Au_4$. At this stage, the consumption rate constant of $(Al,Cu)_3Au_8$ IMC was 6.45×10^{-8} cm/sec$^{0.5}$ and was close to the growth rate

constant of $(Al,Cu)Au_4$ IMC, i.e., 10.0×10^{-8} cm/sec$^{0.5}$. For aging up to 3,000 h[ours], the average thickness of both $(Al,Cu)_3Au_8$ and $(Al,Cu)Au_4$ approached 2.7 μm.

In the edge of wire bonds, the Al-Cu pad was not completely consumed even after 3,000 h[ours of aging. For aging less than 500 h[ours], three IMCs were observed, i.e., $(Al,Cu)Au_4$, $(Al,Cu)_3Au_8$, and $(Al,Cu)Au_2$. After 1,000 h[ours] of aging, $Al_3(Au,Cu)_8$ was revealed between the Al-Cu pad and $(Al,Cu)Au_2$. In addition, $Al_3(Au,Cu)_8$ in the edge of wire bonds gradually grew with aging time.

Editorial Note: That is, Au and Cu act interchangeably with the Al intermetallic compound. The intermetallic compound growth rate with respect to time is Fickian.

In 2007, Karpel, Gur, Atzmon and Kaplan wrote concerning *Microstructural Evolution of Gold-Aluminum Wire-Bonds* [11].

two mechanisms are responsible for the degradation of the bond: an increase of volume by intermetallic formation, which results in stress induced cracks; and oxidation of the interface between the Au and the intermetallics. In order to overcome this failure, an increase of the Au durability to plastic deformation should be achieved.

In 2007, Blish et al. wrote concerning *Gold–Aluminum Intermetallic Formation Kinetics* [12],

"... the "signal" for reliability as a function of Al thickness was weak at best."

"One of the unexpected results was that the IMC growth kinetics and mechanism (the latter inferred from the time exponent) changed with Al thickness. Really thick Al films (3,000 nm) showed an IMC growth time exponent of 1/2, which is characteristic of bulk lattice diffusion, but we saw much smaller time exponents (~1/4) for thin films (470–1,000 nm)."

"... we find that bond lifetime varies (empirically) as a power law on fcc impurity concentration."

"... we looked at the wire analysis (ICP-MS/OES) to see what unexpected element might be responsible for the stellar performance of wire "D." The answer turned out to be arsenic!"

In 2007, Bahi, Lecuyer, Gentil, Fremont, Landesman, Christien and Gall reported their findings in respect to the *Influence of Technological Parameters on the Behavior during Aging at High Temperature of Various Packages, in the Automotive Environment* [13].

"A multi-aromatic resin (Green resin) was used as molding compound; it is an epoxy resin without a flame-retardant system, then Au-Al intermetallic phases are not corroded by the bromine (Br) contained in a flame retardant system. So, the intermetallic thickness grows normally. This failure mechanism is not accelerated by corrosion."

"Although all packages are characterized by the same evolution of intermetallic thickness growth, they didn't present the same thermal behavior. It allows us to conclude that there exists another key factor which influences the thermal aging behavior."

"… the role of the package geometry was identified, and experimental results were confirmed by both simple analytical model and FEM simulations."

In 2007, Choi, Teck, Poh, Hoon and Diong reported [14],

> The breakthrough of semiconductor technology advances at a very fast pace and it challenge to unlimited border. This has resulted in the transition of packaging to be smaller, thinner as well as higher density in circuitry components on single chip as well to much more complicated packaging such SIP, PIP which has more dies or more package stack on each other. This breakthrough is a combination from the creativity of designers/engineers (human), machine, process (method) and materials advancement. This paper will focus on advance materials of Au bonding wire interaction with Al bond pad metallization which have direct impact on packages' performance and reliability.

In 2007, Park, Cha, Kim, Jo, Park, Kim, Shin, Shin, Ouh and Jeon studied *Interfacial Degradation Mechanism of Au/Al and Alloy/Al Bonds under High Temperature Storage Test* with respect to contamination, epoxy molding compound, wire and bonding strength [15].

> Sb comes from the plateau and valley (concave) surfaces of Au–Al bonds encapsulated with the OCN [orthocreosol-novolac], causing an open failure with lifted wire bonds. Certainly, ICP–MS confirms that the OCN is a resource of Br. As results, the degradation mechanism of Au–Al bonds with the OCN between 150 °C and 170 °C is attributed to the interdiffusion of Sb and Br originated from the OCN, leading to compromise mechanical integrity of Au–Al bonds and lowering the activation energy. It is also found that in the case of Au–Al bonds the lifetime with the BP [bisphenol] is much longer than that with the OCN. It manifests that the usage of an alloy wire that can retard the growth kinetics of IMC ensures the long-term reliability of bonding integrity. In consequence, lower flame retardant and higher are critical intrinsic material properties that should be taken into account when a new EMC [epoxy Molding compounds] is introduced to pursue cost effectiveness without losing any reliability issues.

Yeoh Lai Seng, in 2007, presented his work on the *Characterization of Intermetallic Growth for Gold Bonding and Copper Bonding on Aluminum Metallization in Power Transistors* [16].

> Gold wire bonding to aluminum pad[s ha]s [been] a widely applied interconnection method in [the] microelectronics industry for years. Copper wire bonding has gained popularity due to its economic advantage and superior electrical performance [in] compar[ison] to gold wire bonding. It [is well established] that, at elevated temperatures, interdiffusion between [the] ball bond and [the] aluminum pad[s] leads to [the growth of] intermetallic [compounds] and [the] formation [of voids] at the gold-aluminum interface. Our study reveal[ed] that this would influence the drain-to-source on-resistance RDS(on), which is an

important parameter for power MOSFET [metal-oxide-semiconductor field-effect transistor] as it determines how much current the device can carry for low to medium frequency applications. We observe[d] that both gold bonds and copper bonds demonstrate[ed] very stable RDS(on) at 150 °C thermal aging up to a few thousands hours. Nevertheless, at much higher thermal aging temperature of 200 °C, gold bonds start to form Kirkendall voids at early aging stage. The voiding causes the gold ball to be isolated away from the aluminum pad, resulting in undesired drastic rise of RDS(on) value. In contrast, copper bonds render much more stable RDS(on) at 200 °C thermal aging over time owing to the relatively small interdiffusion rate of Cu-Al phases. As a result, copper is more suitable to be utilized as bonding method in MOSFET devices.

Since 2006, Fairchild Semiconductors progressively replaced gold bonding with copper bonding in power transistor products to provide customers with more reliable and longer-lasting devices. In addition, the research findings further reinforced customer confidence in using copper wire bonding devices for high-temperature field applications, based on the key advantage that intermetallic growth defects were not observed in copper bonds at elevated temperatures.
In 2007, Junhui Li, Han Lei, and Zhong Jue presented their findings on *Diffusion on Two Interfaces of Ultrasonic Flip Chip Bonding* [17].

Cross-section features in a flip-chip bonding area were inspected by using a high-resolution transmission electron microscope (HRTEM). For the better flip chip bonding variables, depth of atom diffusion at Au/Al interface of ultrasonic flip chip (FC) bonding was about 500 nm by using a line scanning of HRTEM, and depth of atom diffusion at Au/Ag interface of ultrasonic FC bonding was about 200 nm. If Au/Ag interface received ultrasonic energy first by changing the process of ultrasonic FC, depth of Au/Ag atomic diffusion was about 450 nm, and depth of Au/Al atomic diffusion was about 250 nm. An ultrasonic energy conversion law was discovered that the ratio of first interface to next interface along ultrasonic energy transmitting path in ultrasonic FC bonding was about 2.28:1, which indicated an optimizing process of ultrasonic FC bonding, and a new way of FC bonding was designed for the first interface along the path of ultrasonic energy transmitting.

In 2008, Fernandez presented on *Chemical and Physical Characterization Techniques in Highlighting Intermetallic Compound (IMC) Formation* [18].

Evaluations and qualifications on ball attach process for bump packages, studies on solderability between lead to PCB on leaded packages, wire bond integrity evaluations, these are studies that involves thorough characterization on bonds and intermetallic compound (IMC) layers for different interfaces. For ball attach process in bumped packages, we focus on intermetallic formation between solder ball to the under-bump metal (UBM). For solder integrity, studies focus on the lead post to solder to PCB interface. Intermetallic Compound (IMC) formations between interfaces, is valuable in characterizations. The formation of intermetallics do not just cause good integrity on the interface or bond but can also cause assembly related problems. Example, intermetallics of gold

> and aluminum are a significant cause of wire bond failures in semiconductor devices and other microelectronic devices. In analyzing intermetallic compound formations, there is a need for the IMC layer to be highlighted, for interface layer difference, for measurement on the thickness, and to merely highlight. This paper aims to present a study on the manual techniques in highlighting intermetallics. Chemical etching, delayering, and immersion on chemical solutions, done on different types of samples to highlight intermetallic compound (IMC).

In 2008, Sritharan, Li, Xu and Zhang published the results of their study on *Oxidation of Al–Au Intermetallics and its Consequences Studied by X-ray Photoelectron Spectroscopy* [19].

> The degree of oxidation increases in the order Al_2Au <$AlAu_2$ < $AlAu_4$. The oxide layer thickness in Al_2Au is estimated to be 10 nm. $AlAu_4$ has the greatest tendency to be oxidized among the intermetallics tested, suggesting that the oxidation of $AlAu_4$ in wire bond could be a serious reliability problem.

Editorial Note: The abstract appears to conflate "oxidation" with intermetallic layer thickness and the extent to which the phases are enriched in gold.

In 2008, Manoubi Auguste Bahi et al. published their study on the *Degradation Mechanisms of Au–Al Wire Bonds during Qualification Tests at High Temperature for Automotive Applications: Quality Improvement by Process Modification* [20].

> the use of Au–1 wt.% Pd instead of pure Au is investigated, showing that this alloy permits to decrease the Au–Al intermetallic growth and, hence, to increase the lifetime.

In 2008, Bahi et al. presented *Bond Reliability Improvement at High Temperature by Pd Addition on Au Bonding Wires* [21].

> To extend the qualification domain to the biggest package [market] size in the automotive environment at high temperature, it is necessary to improve the Au-Al bond reliability. Thus, the improvement provided by addition of palladium as alloying elements to Au bonding wires is evaluated. In this study, the use of Au-1 wt%Pd instead of pure Au is investigated; samples are aged at 175 and 200 °C. In the case of the Au-1 wt%Pd wire, a Pd-rich layer was found at Au/Au-Al interface, which prevented Au diffusion and yielded an even slower intermetallic (IMC) growth rate at the Au-Al interface as compared to the case of the pure Au wire. Wire pull test after thermal aging proved that Pd addition enhanced bond reliability and increased the life-time. Cross-sectional SEM and EDX analyses (mapping and linescan extraction) allowed to identify Au-Al IMC phases and to understand the role of Pd rich layer at the Au/IMC interface. Investigations on the failure mechanisms provided an explanation for the improved robustness of the connections in the case of the alloyed wires as compared to in purse gold wires.

Editorial Note: It is unclear why Pd would impede diffusion, given that the Pd atom is smaller than both Au and Al, and all three elements have an fcc crystal structure.

However, the Pd-Au phase diagram indicates the formation of metastable AuPd around 200–300 °C, along with Au_3Pd and $AuPd_3$ phases at 1,000–1,100 °C, the latter two being irrelevant in this context

Akira Furuto and Masanori Kajihara published *Numerical Analysis for Kinetics of Reactive Diffusion Controlled by Boundary and Volume Diffusion in a Hypothetical Binary System* in 2008 [22].

> When the size of the basal-plane remains constant independently of the annealing time, the thickness of the compound layer is proportional to the square root of the annealing time.

In the same year 2008, Jalar, Rosle and M. Hamid published *Effects of Thermal Aging on Intermetallic Compounds and Voids Formation in Au-Al Wire Bonding* [23].
AuAl intermetallic compounds continue to grow with increasing temperature and time. Initial degradation [is] activated by the rapid formation of Kirkendall voids at the Au-AuAl interface [intermetallic thickness]. The intermetallic layer thickness was subsequently observed to [… increase over the time interval …] from 500 hours to 1,000 hours at 150 °C. The decrease in shear strength after 1,000 hours of aging can be attributed to the severe formation of Kirkendall voids, as confirmed by microstructural analysis
Pufall, Alpern, Kanert, Pfost, Smorodin and Stecher presented their study on *Wire Bonding Degradation Induced by Temperature Gradients under Active Cyclic Loading* in 2009 [24].

> It is well known that high temperature storage can degrade wire bonding contacts significantly due to interdiffusion of pad metal and bonding wire. Looking at harsh applications such as engine management we notice an additional failure mode caused by the temperature gradient during the pulsed active cycling period. Especially when we aim at components with high temperature capability and we substitute the power aluminum metallization with power copper in order to avoid the formation of lifetime limiting intermetallics, the degradation of wire bonds (Au, Al, Cu) must be assessed with respect to the electrical pulse width, the dissipated power and the number of active cycles, which can exceed 500 million [cycles] in automotive applications. This paper presents experimental data with different temperature stress. The time dependent temperature distribution in the device is determined with an electro-thermal simulator (TESI). The calculated temperature gradients will be used to enable a thermal-mechanical simulation (ANSYS). As a result, a prediction, which kind of pulses can reduce the lifetime of the components under investigation, should be possible.

In 2009, Schneider-Ramelow, Schmitz, Schuch and Grubl studied *Kirkendall Voiding in Au Ball Bond Interconnects on Al Chip Metallization in the Temperature Range from 100–200°C after Optimized Intermetallic Coverage* [25].

> The presentation addresses the reliability of Au ball bonds of different Au wire qualities on Al chip metallization of different thicknesses and compositions at

> temperature storage from 100 to 200 °C up to 4,000 h[ours]. In this context the interfacial reactions and intermetallic phase coverage directly after the bonding process was optimized to get the best starting condition for phase growth at elevated temperatures and to avoid critical Kirkendall void growth. This failure mechanism is influenced by numerous factors, such as aging temperature and time, Au wire and Al metallization composition and ratio of mixture as well as the percentage area of interconnection formation under the ball. These influences are mainly responsible for ball lift offs under operating conditions. In many cases lift offs already occur at Al metallization thicknesses > 1 µm and temperatures in the range of 175 °C, while temperatures up to 150 °C or at 200 °C are less critical. Investigations include mechanical tests of Au loops and ball contacts as well as microstructure observations of the contacts in correlation to material composition, aging temperature and Al metallization thickness. Au/Al intermetallic phase thicknesses below the Au contacts on Al metallization are typically a few hundred nanometers thick directly after the bonding process, depending on bonding conditions like process parameters and material combination. These phases grow under temperature influence and Kirkendall voiding can occur. A most significant result in this context is that pull and shear lift offs occur if the chip metallization is clearly thicker than 1 µm and intermetallic phase coverage (after bonding) is less than 2/3 of the bottom side ball area. These results will considerably contribute to a better understanding of Kirkendall voiding failure mechanisms.

In 2009, Breach and Wulff published about *Intermetallic Growth in Gold Ball Bonds Aged at 175 °C: Comparison between Two 4N Wires of Different Chemistry* [26].

> Two wires of different chemistry type a and type b, ballbonded to the identical bondpad metallization grow the same intermetallics during parabolic growth but after extended ageing Au_8Al_3 undergoes a phase transformation into a compound, which from EDX analysis, has the same composition as Au_4Al... The occurrence of the phase transformation only in type b wire only is attributed to dopants in the wire that may interact with other dopants and/or with bondpad elements and reduce the activity of Au_8Al_3. The resulting activity gradient is proposed to drive the diffusion of gold from the wire into the Au_8Al_3 causing the transformation of Au_8Al_3 into Au_4Al.

In 2009, Bahi, Fremont, Landesman, Gentil and Lecuyer presented *a New Methodology for the Identification of Ball Bond Degradation during High-temperature Aging Tests on Devices in Standard Plastic Packages* [27].

> We describe a new methodology for the "in-situ" identification of wire-bond degradation at early stages during high-temperature aging tests on devices with standard plastic packages. This methodology is based on the measurement of the changes in wire bond resistance, which is deduced from the I(V) characteristics of the ESD protection diodes on each contact pad of the circuit.

In 2009, Maldonado, Oliva and Corona studied the effects of annealing on thin AuAl films [28].

> Abrupt changes in rms-roughness (grain size) values produce inverse changes in the electrical resistivity. By changing Au atomic concentration we found higher electrical resistivities for 40 % and 60 % Au concentrations. However, the electrical resistivity remains constant for 50 % Au atomic concentration. By changing the alloy thickness the electrical resistivity increases with the decrement of bimaterial thickness, a typical behavior for thin films scale. Only for nanofilms (thickness below 100 nm) the electrical resistivity increases drastically.

Bahi et al. presented a new diagnostic methodology for package delamination as an indicator of ball bond lift in 2009 [29].

> The components used in the automotive environment are subjected during their lives to various environmental combined stresses. The predominant failure mechanism at the package level due to these combined stresses is the ball bond lift. The failure is due to the combination of metallurgical effects, such as the intermetallic (IMC) Au-Al thickness growth, widening of Kirkendall voids and mechanical ones as a consequence of the delamination increase. The purpose was to detect the potential weaknesses of die to leads connections. The usual methods described by the automotive qualification standard AEC-Q100 are not sufficient to follow the temporal evolution of the degradation. For this reason, we propose, in addition to the existing methods, a new diagnosis reliability methodology, which consists to use the delamination monitoring as an indicator of potential assembly weaknesses.

In 2009, Shin, Song, Park, Kwon, Moon, Cho, Yoo, Byun, Joh, Do and Paik analyzed the interfacial failures of ultrafine pitch wires with low intermetallic coverage [30].

> Ultra-fine pitch Au wire has become inevitable in the electronic packaging industry due to decreasing electronic package dimension and increasing cost of gold. However, fine pitch Au wire with diameter less than 0.5 mil. (12 μm), bonded on Al pad, showed low inter-metallic coverage, and it caused distinctive interfacial failures: the oxidation of Au_4Al and overgrowth of inter-metallic compound (IMC). In order to enhance the interfacial failure, addition of Pd was proposed due to its known effect as the diffusion barrier. In this study, 0 ppm, 3,000 ppm, 6,000 ppm, and 10,000 ppm of Pd was added to 0.5 mil. Au wire as alloying element and subsequently aged under high temperature storage test (HTST) in air and vacuum condition at 175 °C. By the observation of Au/Al interface using scanning electron microscope (SEM) and reliability test using ball pull test (BPT), it was confirmed that major failure mode was the oxidation of Au_4Al for air-aging and overgrowth of IMC for vacuum-aging; however, it was observed that the interfacial failures were affected by the addition of Pd. For air-aging, the oxidation rate was in order of 10,000 ppm (fastest) > 0 ppm = 6,000 ppm > 3,000 ppm (slowest) of Pd showing that the interfacial failure due to the oxidation was affected by Pd contents. For vacuum-aging, higher Pd contents decreased overgrowth of IMC and number of ball-lift. Detailed effects of Pd for the oxidation rate were analyzed using transmission electron microscope (TEM). Au wire with 2,500 ppm of Pd was analyzed as most oxide-stable Pd composition, and was observed to form segregation of Pd

> mass at the grain boundary of Au_4Al, and the segregation protected Au_4Al from the oxidation. However, the sample with 10,000 ppm of Pd made separate Pd-rich grain, and it didn't prevent the oxidation and even caused quicker oxidation than the sample without Pd.

Editorial Note: Intermetallic formation is **not** oxidation.

In 2009, Xu, Liu, Silberschmidt, Chen and Sivakumar presented their findings on *TEM Interfacial Characteristics of Thermosonic Gold Wire Bonding on Aluminum Metallization* [31].

> A native alumina thin film existing on aluminum and its evolution in thermosonic gold wire bonding were characterized using high resolution transmission electron microscopy (HR TEM). It has been found that the partly fracture of this alumina film by the ultrasonic vibration allows the interdiffusion at the Au/Al interface, such that the intermetallic compounds (IMCs) are formed. The IMCs are identified as Au_4Al and $AuAl_2$ with a thickness of approximately 300 nm. With a higher ultrasonic energy, IMCs are formed in more contact areas, resulting in a stronger bond. Based on these experimental results, a mechanism of ultrasonic gold wire bonding is proposed, which helps to understand the correlation among ultrasonic energy, interfacial structure and bonding strength.

Editorial Note: Interesting that they can get TEM to see fractures in an oxide film only 2 layers deep. This supports previous findings suggesting that ultrasonic energy does not cause heating at the interface

In 2009, Tok and Effie presented their work on *Ultra-fine-wire Applications and Challenges in 30 μm Process* [32].

In UFP bonding, the use of 2N wires is required due to their reliability and robustness. However, 2N wires gives exhibit a significantly higher occurrence of bonded ball non-concentricity compared to than their 3N and 4N counterparts. This non-concentricity poses a potential risk to long-term reliability, as the "off-centered" balls often result in non-uniform intermetallic compound formation beneath the bond, promoting the development of large voids over time during aging.

In 2010, McCracken, Kim, Mayer, Persic, Hwang and Moon presented their report on *Assessing Au-Al Wire Bond Reliability using Integrated Stress Sensors* [33].

> Wire bond reliability testing typically consists of aging bonds in a high temperature environment for long time periods, removing samples at intervals to assess bond shear strength and characterize the bond cross sections. In this way, the degradation of the bond can be monitored at discrete time intervals, and it is determined whether the bond will be reliable under long term operation at lower temperatures. This process is labor and time consuming. An alternative process is reported using piezoresistive integrated CMOS microsensors located around test bond pads. The sensors are sensitive to radial compressive or tensile stresses occurring on the bond pad due to intermetallic formation, voiding, and crack formation at the bond interface. Two sets of identical test chips are bonded with optimized Au ball bonds and aged for 2,000 h[ours] at 175 °C. One set is connected to equipment which monitors signals from the stress sensors, along

with the contact resistance of the bonds. The other set of chips is destructively tested by shear tests and cross sectioning. It is found that the stress sensors are capable of indicating which stage of intermetallic growth is currently being experienced, by relating the signal to the relative density of the intermetallic compounds (IMCs) which form during aging. The sensors can detect the consumption by IMCs of each Al layer in a multilayer pad and can monitor the formation of $AuAl_2$ which indicates an advanced stage of aging. Sensor signals combined with contact resistance measurements provide a valuable tool for preliminary reliability studies and give real-time insight into microstructural changes. Drop in shear strength of a ball bond is detected by a change in the microsensor signal combined with a contact resistance increase.

In 2010, Krause, Mäŕz, Dresbach and Petzold presented *Electron Backscatter Diffraction Microstructure Investigations of Electronic Materials down to the Nanoscale* [34].

In this paper, the application of electron backscatter diffraction (EBSD) methods to materials used for common microelectronic interconnection technologies is demonstrated with particular emphasis to lead-free soldered interfaces and thermosonic wire bond interconnects. Here, the paper focuses on the quantitative analysis of grain orientation, grain size and grain distribution of SAC as well as gold, aluminum, and copper bonding wire materials. In addition, special attention is paid to high resolution analysis of the intermetallics formed in the interfaces of microelectronic packaging interconnects. The application and the potential of EBSD to detect the Cu/Sn, Ni/Sn and Au/Al intermetallics being practically relevant for soldering and wire bonding is demonstrated.

In 2010, Krause, Petzold, März and Bennemann presented *High resolution analysis of intermetallic compounds in Microelectronic Interconnects using Electron Backscatter Diffraction and Transmission Electron Microscopy* [35].
While EBSD-based techniques can provide valuable support, they cannot fully replace detailed TEM studies due to their limited spatial resolution and the superior capability of TEM to analyze interface defects and nanometer-sized voids.
In 2010, März, Scheibe, Graff and Petzold presented *Growth Behaviour of Gold-Aluminum Intermetallic Phases (IMP) in Temperature Aged Ball Bonds observed by Electron Backscatter Diffraction* [36],

Gold wire bonding on aluminum pads plays still an important role in microelectronic packaging. The lifetime of these bonds is mostly determined by the intermetallic phases (IMP) at the gold/aluminum interface. To investigate the phase growth after the wire bonding process, bond samples were produced. Several samples have been exposed to elevated temperatures to initiate further phase growth. To obtain different reaction stages of the intermetallic compound formation, the samples have been annealed for different defined times in a furnace. As an alternative to common methods like transmission electron microscopy (TEM), the electron backscatter diffraction (EBSD) technique has been applied for phase identification in the Au/Al system. After optimization of the cross-section preparation as well as the internal parameters of the EBSD software, EBSD phase and orientation mappings of bond interfaces have been conducted. It has been shown that the different IMPs are reliably distinguishable by their crystallographic structure. In addition, the transformation of the phase regions within the interconnect at different reaction stages could be determined.

In 2010, Xu, Silberschmidt, Pramana, White, Chen, Sivakumar and Acoff published *a Micromechanism Study of Thermosonic Gold Wire Bonding on Aluminum Pad* [37].

Au_8Al_3 is the correct composition and structure determined by Range and Buechler … but first identified by Francombe *et al.* Using electron diffraction as Au_5Al_2 (Au:Al=2.5). While there are frequent literature references to Au_5Al_2 there is, as yet, no definitive proof of its existence. Pretorius *et al.* … assumed Au_5Al_2 for determining ΔH.

Editorial Note: Each line (K or L, etc.) has multiple energies.

The thermodynamics and kinetics of Au–Al bond formation are difficult to evaluate in situ and poorly understood. If IMC growth during annealing is diffusion-limited it follows a parabolic law … such that,

$$x = (Dt)^{1/2} \tag{1}$$

"…and…"

$$D = D_0 \exp(-Q/RT), \tag{2}$$

TABLE 12.1
Crystallographic data for compounds and polymorphs of the Au–Al system (Notes have been omitted)

Chemistry		Crystallochemical parameters			
Composition[a]	Au:Al	Polymorph/ compound symmetry	Space group	Atoms per unit cell	Unit cell volume (Å^3)
$Au_{23}Al$	23	Cubic	$Fm\bar{3}m$	4	67.28[c]
$Au_{86}Al_{14}$	6.14	Cubic	$Fm\bar{3}m$	4	66.85[c]
Au_4Al	4	Cubic (β')	$P2_13$	20	331.76[c,d]
		Cubic (β)	$Im\bar{3}m$	2	34.01[e,f]
		Tetragonal	$I4/mmm$	2	33.02[c,g]
Au_8Al_3 [h]	2.66	Rhombohedral	$R\bar{3}c$	44	724.77[i,f]
Au_2Al	2	Orthorhombic	$Pnma$	12	190.54[c,j]
		Orthorhombic	$Pnnm$	30	475.16[c,k]
AuAl	1	Cubic	$Pm\bar{3}m$	2	30.96[e]
$AuAl_2$		Monoclinic	$P2_1/m$	32	135.26[i]
	0.5	Cubic	$Fm\bar{3}m$	12	215.71[c]

where x is the IMC thickness at time t, D the growth rate constant, D_0 a prefactor, Q the activation energy, R the molar gas constant, and T the absolute temperature.

1) A compact amorphous alumina layer 5 nm thick is present in aluminum prior to bonding and acts as a barrier to Au and Al interdiffusion.
2) Ultrasonic vibration partially disrupts the continuity of the alumina film to create favorable diffusion pathways that promote the local formation of IMCs. At higher ultrasonic energies, alumina fragmentation is greater, metal mobility accelerates, and IMCs more readily form in the contact area, resulting in a stronger bond. The dominant alloys are Au_4Al and $AuAl_2$ (total thickness of 150–300 nm), with remnant alumina encapsulated in these IMCs.
3) With the simplifying assumption that IMC formation during bonding is diffusion controlled, we estimate the ultrasonic effect in the current work to be equivalent to heating the interface to 600 °C, while the overall average temperature is probably near 320 °C. This arises because ultrasonic vibration increases the real local temperature and simultaneously produces numerous diffusion pathways via grain boundaries and dislocations that lower activation energies.
4) The coexistence of Au_4Al and $AuAl_2$ is due to a native alumina overlayer that attenuates Au and Al interdiffusion. This reduces the effective concentration of Au at the growth interface adjacent to Al and that of Al at the interface near Au.

In 2010, Jiang, Sun, Wang, Min and Chen presented the results of their *Study of a Practicable Wire Bonding Method for Applying Copper Wire Bond to Large-scale Integrated Circuits* [38].

> micro-structure analysis focused on IMC change at two interfaces of Au-Al and Cu-Al couples through the reliability tests. The study result finally indicates that COG [copper-on-gold] bonding is a practicable approach to achieve copper wire application to large-scale integrated circuits with risk-free.

In 2010, Yingwei Jiang, Sun, Yu, Wang and Chen published the *Formation and Growth of Intermetallic Compounds in Cu–Au and Au–Al Systems for Copper on Gold Bonding* [39].

> Compared with Au wire bonding, a comparable Au–Al IMC formation and growth is shown in the COG bonding. The Au–Al IMC formation and bonding strength between Au bump and Al metallization are mainly related to bonding parameter setting for Au bump.

In 2010, Beleran, Hong, Wilson, Zhang, Amurao, Mehta and Zhu presented *Solutions for 45/40nm ELK Device Integration into Flip Chip and Wire Bond Packaging* [40]. Many semiconductor manufacturers are choosing 2N wire for ultra-fine-pitch gold wire applications, this is not just because of the more cost-effective copper wire, but primarily because 2N provides better ball bond reliability compared to conventional 4N. Many recent studies have shown that with the addition of dopants like Pd, Pt and Cu help retard voids formation as well as oxidation. These dopants play an important role in inhibiting Au diffusion into the intermetallic and thereby reduces Al precipitation and voids formations.

After an unbiased highly accelerated stress test (HAST) (condition: 130 °C, 85 % Relative Humidity) for 96 hours, 4N Au and Cu wire type have shown leakage failures. But they did not see any leakage failures for 2N gold wire, therefore,

> 2N gold wire type is now the preferred material wire type especially for fine and ultra fine pitch packages where conventional 4N has difficulty achieving the required stringent reliability performance of microelectronic packages.

In 2011, Minho O and Masanori Kajihara reported the results of their study of the *Kinetics of Solid-State Reactive Diffusion between Au and Al* [41].

> In order to examine the growth behavior of the Au-Al compounds, the kinetics of the reactive diffusion was experimentally observed using sandwich Al/Au/Al diffusion couples. The diffusion couple was prepared by a diffusion bonding technique and then isothermally annealed at temperatures of 623–723 °K for various times up to 103 h. Under such annealing conditions, the diffusion couple is practically considered semi-infinite. Owing to annealing, compound layers consisting of $AuAl_2$, AuAl and Au_8Al_3 are produced at the initial Au/Al interface in the diffusion couple. The volume fraction in the compound layers is larger for Au_8Al_3 than for AuAl and $AuAl_2$ in the early stages but becomes smaller for Au_8Al_3 than for AuAl and $AuAl_2$ in the late stages. However, there are no systematic dependencies of the volume fraction on the annealing time and temperature. The total thickness of the compound layers is proportional to a power function of the annealing time. The exponent of the power function is smaller than 0.5 at 673–723 °K. The exponent smaller than 0.5 means that the interdiffusion across the compound layers governs the layer growth and boundary diffusion predominantly contributes to the interdiffusion. On the other hand, at 623 °K, the exponent is equal to unity for annealing times shorter than 8 hours but smaller than 0.5 for those longer than 8 hours. The exponent of unity indicates that the layer growth is controlled by the interface reaction at the migrating interface. Thus, at T = 623 °K, the transition of the rate controlling process occurs at a critical annealing time of 8 h[ours].

Editorial Note: Intermetallic compound phases observed would depend on initial thicknesses of Au and Al.

In 2011, Breach and Lee presented their *Conjecture on the Chemical Stability and Corrosion Resistance of Cu-Al and Au-Al Intermetallics in Ball Bonds* [42].

> Under moist conditions and in the presence of chlorine (Cl), copper ball bonds appear more prone to corrosion than gold ball bonds. One reason for this may be that Cu-Al intermetallic compounds are more susceptible to attack by moisture than Au-Al intermetallic compounds. The presence of Cl may also accelerate this effect. This article suggests that the effective heat of formation of Cu-Al

and Au-Al compounds, which is a measure of stability and chemical resistance to change, may be correlated with surface energy of the various compounds. Due to lack of experimental data, surface energy of compounds is calculated in the liquid state and an assumption is made that the difference between liquid and solid compound surface energies is negligible. An initial conjecture is made that low surface energy compounds are more stable than those with high surface energy. Results of calculations for Cu-Al and Au-Al compounds suggest that $CuAl_2$ may be more stable than Cu_9Al_4 due to the latter having a smaller effective heat of formation (EHOF) and a higher surface energy. This is in line with experimental observations; However, the initial conjecture is modified by including work function because of the increase of surface energy of Au-Al compounds is accompanied by high stability. A further conjecture is that there is a relationship between work function, corrosion resistance and EHOF.

In 2011, Yao, Lai, Liu and Liang presented *the Effect of Thermal Stress on High Density Packaging Integrated Circuits* [43],

In this paper, an accelerated thermal reliability test method for evaluating the packaging reliability of ICs which includes hot step, cold step and rapid thermal cycling test is presented. The technology of FIMV (Force current measure voltage) was applied during the reliability test as an indicator of degradation of packaging property, which allowed the reliability performance of ICs to be assessed in real time. The experimental results showed that the thermal stress resulted in the degradation of interfacial adhesion of plastic packaging ICs. Because of the temperature changing during the rapid thermal test, the strain and stress due to the coefficient of thermal expansion (CTE) mismatch between the encapsulant and the adjacent materials could contribute to delamination or de-adhesion. In some cases it was directly linked to a failure if some severe defects occurred because of delamination, such as wire bond lift-off or fracture. Crack in die attach adhesive based on the same failure mechanism was also found. Additionally, unwanted brittle Au-Al intermetallic compound was detected at the bond interface because of the effect of high temperature. The formation of the Au-Al intermetallic compound led to the increase of electrical resistance and the weakening of bond strength which resulted in bond lift-off finally. At last future research work in this field is suggested.

Bin Yao and Ping Lai presented *Accelerated Reliability Evaluation for High Density Packaging Integrated Circuits* in 2011 [44].

Differing from the traditional reliability prediction methods, the technology of FIMV (Force current, measure voltage) was applied during the reliability test which allowed the reliability performance of ICs to be assessed in real time. It was found that the voltage changed correspondingly for the different stress changing conditions. So, it is validated that the use of FIMV technology as an indicator of degradation of packaging property is effective. The experimental results show that thermal cycling stress can affect the adhesion of the interfaces in plastic packaging ICs. Temperature change during the thermal cycling test and the mismatch between the CTEs of the encapsulant and the adjacent material of plastic packaging ICs can lead to thermo-mechanical stresses that can cause delamination or de-adhesion. Additionally, it is found that unwanted Au-Al intermetallic compound may form at the bond interface because of the effect of high temperature. The formation of the unwanted Au-Al intermetallic compound will reduce the bond strength and lead to wire bond lift-off finally. At last future research work in this field is suggested.

Dechun Lv and Guangyu Yang presented *Gold Bonding Reliability in High Density FPGA Devices* in 2011 [45].

The formation of intermetallic mass along the edge of the ball has been observed inherent to the gold bonded Aluminum pads. In a well-bonded sample, it occurs only near the edge of the ball resulting to still strong, reliable bonds after the parts have been exposed at a long duration to elevated temperature. Hence, the presence of intermetallic mass along the periphery of the ball is not an indication of unreliable bond.

In 2011, Xu et al. presented their observations on *Void Growth in Thermosonic Copper/Gold Wire Bonding on Aluminum Pads* [46].

Void growth in wire bonds significantly influences the reliability of electronic devices. This paper found that, in the as-bonded state, a few voids nucleate in Au-Al bonds, while they are absent in Cu-Al bonds. Voids grow much faster in Au-Al bonds than in Cu-Al bonds during thermal annealing. It is proposed that void growth in Au-Al bonds due to the oxidation of IMCs and volumetric shrinkage resulted from the growth of Au_8Al_3 and Au_4Al, but not due to the Kirkendall effect. Large voids up to 10 μm exist after extended annealing, which is also attributed to the outward diffusion of Au to react with Al in the area beyond the perimeter of the bonds. In Cu-Al bonds, the void growth rate is low, and only a few voids of <100 nm are present along the alumina remnant after prolonged annealing at a high temperature (e.g., 250 °C for 25 h[ours]). The formation of void in Cu-Al bonds is probably due to the volumetric shrinkage during the formation of Cu_9Al_4. The oxidation of Cu-Al IMCs and outward diffusion of Cu is insignificant; therefore, the void growth rate in Cu-Al bonds is low.

In 2012, Delucca, Osenbach and Baiocchi presented their observations of intermetallics formation for Au-Al wire bonds [47].

With heat treatment, even simply that from package processing, the IMC microstructure quickly takes on a stratified thin/thick morphology that persists with aging and phase changes. The interface between the thin and thick IMC is observed at the location where greater voiding/low-density interface development takes place, as opposed to the Au/IMC interface. Pd accumulates in the IMC that resides at the Au/IMC interface for 2N wire bonds and appears responsible for a phase evolution different from that for 4N wire.

In 2012, Huang, Tan, Walter, Pape, Fan and Koerner presented their results of the *Simulation of Diffusion Controlled Intermetallic Formation at Au/Al Interface* [48].

> This paper describes the finite element simulation for diffusion controlled intermetallic formation of Au/Al interface during the wire bonding process. The analogous correlation between the intermetallic atomic diffusion and the heat transport is introduced, and the simulation of intermetallic diffusion using a commercial finite element method (FEM) tool ANSYS is demonstrated. Phase dependent diffusivities have been implemented into the material model in order to more accurately model the distribution of diffusing atomic concentration. The various phases of intermetallic compounds (IMCs) can be determined briefly by the obtained atomic concentration profile from FEM and the phase composition table. The overall thickness of the IMCs layer can then be calculated. In this paper, the concept for the subsequent stress analysis based on the volume change due to diffusion is also discussed.

In 2012, Breach and Lee published the results of a comparison *of Shear Strength and Failure Modes of As-Bonded Gold and Copper Ball Bonds on Aluminum Metallization* [49].

> Copper and gold ball bonds on aluminum metallization, bonded under optimum conditions that maximize intermetallic coverage, normally show fundamentally different failure modes during shear testing. Gold ball bonds are relatively soft compared with the bond pad, and the shear tool removes the upper part of the ball, leaving the lower part of the ball adhered to the bond pad. Copper balls are generally harder than gold and are sheared off the soft aluminum metallization with failure occurring by plastic deformation in the soft bond pad.

In 2019, Ahmad and Smith published results of their Au/Al wire bond interface resistance degradation rate simulations, and concluded [50],
The Au-Al bond degradation in semiconductor applications can be divided into three phases. Resistance increase in the first phase of degradation is purely due to Au-Al diffusion and follows the $t^{1/2}$ law. Resistance increase in the second and third phases is dominated by the rate of formation of Kirkendall voids. This paper shows why the resistance increase, and thus the failure rate, in the final two phases of Au-Al wire bond degradation may not follow the $t^{1/2}$ law for diffusion-driven phenomenon and may show non-Fickian resistance increase despite the degradation being driven by diffusion.

REFERENCES

[1] S.-A. Gam, H.-J. Kim, J.-S. Cho, Y.-J. Park, J.-T. Moon and K.-W. Paik, "Effects of Cu and Pd Addition on Au Bonding Wire/Al Pad Interfacial Reactions and Bond Reliability," *Journal of Electronic Materials,* vol. 35, no. 11, pp. 2048–2055, 2006.

[2] R. Blish, S. Li, H. Kinoshita, S. Morgan and A. Myers, "Gold-Aluminum Intermetallic Formation Kinetics," in *Reliability Physics Symposium Proceedings, 2006. 44th Annual., IEEE International*, San Jose, 26–30 March 2006.

[3] F. W. Wulff, C. W. Tok and C. D. Breach, "Oxidation of Au_4Al in Gold Ball Bonds," *Materials Letters,* vol. 61, no. 2 (January), pp. 452–456, 2007.

[4] R. Klengel, H. Heideallee Knoll, M. Petzold, M. Wohnig and L. Schrapler, "Intermetallic Compound Formation in Au/Al Thermosonic Wire Bonding during High Temperature Annealing at 150 °C as a Function of Wire Material," in *Electronics System Integration Technology Conference, 2006. 1st. Volume 1*, 5–7 September 2006.

[5] L. Hung, Y. Lin, S. Chn, Y. Wang and C. Hsiao, "The Characterization of Intermetallic Growth in Copper and Gold Ball Bonded on Thicker Aluminum," in *Electronic Packaging Technology, 2006. ICEPT '06. 7th International Conference on*, 26–29 August 2006.

[6] T. Muller, L. Schrapler, F. Altmann, H. Knoll and M. Petzold, "Influence of Intermetallic Phases on Reliability in Thermosonic Au-Al Wire Bonding," in *Electronics Systemintegration Technology Conference, 2006. 1st. 5-7 Sept. 2006 Volume: 2*, 2006.

[7] M. Li, S. Li and H. Ji, "Metallurgical Behavior of Au/Al Bond Interface during High Temperature Storage of Ultrasonic Wedge Bonding," in *Electronic Packaging Technology, 2006. ICEPT '06. 7th International Conference on,*, 26–29 August 2006.

[8] Z. Xu and T. Suga, "Room/Low Temperature Interconnection Technique on Micro-bump/Film for COC and COF System," in *Electronic Packaging Technology, 2006. ICEPT '06. 7th International Conference on*, 26–29 August 2006.

[9] A. Teverovsky, "Effect of High Temperature Storage in Vacuum, Air, and Humid Conditions on Degradation of Gold Aluminum Wire Bonds in PEMs," NASA Goddard Space Flight Center, pp. 1–8, 2006.

[10] G.-Y. Jang, J.-G. Duh, H. Takahashi and D. Su, "Solid-State Reaction in an Au Wire Connection with an Al-Cu Pad During Aging," *Journal of Electronic Materials,* vol. 35, no. 2, pp. 223–232, 2006.

[11] A. Karpel, G. Gur, Z. Atzmon and W. D. Kaplan, "Microstructural Evolution of Gold-Aluminium Wire-Bonds," *Journal of Material Science,* vol. 42, pp. 2347–2357, 2007.

[12] R. C. Blish, S. Li, H. Kinoshita, S. Morgan and A. F. Myers, "Gold–Aluminum Intermetallic Formation Kinetics," *IEEE Transactions on Device and Materials Reliability,* vol. 7, no. 1, pp. 51–63, March 2007.

[13] M. A. Bahi, P. Lecuyer, A. Gentil, H. Fremont, J.-P. Landesman, F. Christien and R. L. Gall, "Influence of Technological Parameters on the Behavior during Aging at High Temperature of Various Packages, in the Automotive Environment," in *Proceedings of the 14th IPFA 2007*, Bangalore, India, 2007.

[14] Y. C. Choi, S. C. Teck, L. T. Poh, H. P. Hoon and S. Diong, "Towards Higher Reliability Challenge of Au-Al intermetalic system," in *Electronics Manufacturing and Technology, 31st International Conference on*, 8–10 November 2007.

[15] J. Park, H.-J. Cha, B.-S. Kim, Y.-B. Jo, J.-K. Park, S.-Y. Kim, S.-C. Shin, M.-Y. Shin, K.-I. Ouh and H. Jeon, "Interfacial Degradation Mechanism of Au/Al and Alloy/Al Bonds under High Temperature Storage Test: Contamination, Epoxy Molding Compound, Wire and Bonding Strength," *IEEE Transactions on Components and Packaging Technologies,* vol. 30, no. 4 (December), pp. 731–744, 2007.

[16] Y. L. Seng, "Characterization of Intermetallic Growth for Gold Bonding and Copper Bonding on Aluminum Metallization in Power Transistors," in *IEEE 2007 9th Electronics Packaging Technology Conference*, Singapore, 2007.

[17] J. Li, H. Lei and Z. Jue, "Diffusion on Two Interfaces of Ultrasonic Flip Chip Bonding," in *High Density Packaging and Microsystem Integration, 2007. HDP '07. International Symposium on*, 26–28 June 2007.

[18] J. C. M. Fernandez, "Chemical and Physical Characterization Techniques in Highlighting Intermetallic Compound (IMC) Formation," in *Physical and Failure Analysis of Integrated Circuits, 2008. IPFA 2008. 15th International Symposium on the*, 7–11 July 2008.

[19] T. Sritharan, Y. Li, C. Xu and S. Zhang, "Oxidation of Al–Au Intermetallics and its Consequences Studied by X-ray Photoelectron Spectroscopy," *Journal of Material Research,* vol. 23, no. 5 (May) , pp. 1371–1382, 2008.
[20] M. A. Bahi, P. Lecuyer, A. Gentil, H. Fremont, J.-P. Landesman, F. Christien and R. L. Gall, "Degradation Mechanisms of Au–Al Wire Bonds during Qualification Tests at High Temperature for Automotive Applications: Quality Improvement by Process Modification," *IEEE Transactions on Device and Materials Reliability,* vol. 8, no. 3 (September 2008), pp. 484–489, 2008.
[21] M. A. Bahi, P. Lecuyer, A. Gentil, H. Fremont, J.-P. Landesman and F. Christien, "Bond Reliability Improvement at High Temperature by Pd Addition on Au Bonding Wires," in *10th Electronics Packaging Technology Conference, 2008. EPTC 2008*, 9–12 December 2008.
[22] A. Furuto and M. Kajihara, "Numerical Analysis for Kinetics of Reactive Diffusion Controlled by Boundary and Volume Diffusion in a Hypothetical Binary System," *Materials Transactions,* vol. 49, no. 2, pp. 294–303, 2008.
[23] A. Jalar, M. F. Rosle and M. A. A. Hamid, "Effects of Thermal Aging on Intermetallic Compounds and Voids Formation in AuAl Wire Bonding," *Solid State and Technology,* vol. 16, no. 2, pp. 240–246, 2008.
[24] R. Pufall, P. Alpern, W. Kanert, M. Pfost, T. Smorodin and M. Stecher, "Wire Bonding Degradation Induced by Temperature Gradients under Active Cyclic Loading," in *Thermal, Mechanical and Multi-Physics Simulation and Experiments in Microelectronics and Microsystems, 2009. EuroSimE 2009. 10th International Conference on*, 29 April 2009.
[25] M. Schneider-Ramelow, S. Schmitz, B. Schuch and W. Grubl, "Kirkendall Voiding in Au Ball Bond Interconnects on Al Chip Metallization in the Temperature Range from 100–200 °C after Optimized Intermetallic coverage," in *Microelectronics and Packaging Conference, 2009. EMPC 2009. European.*, 15–18 June 2009.
[26] C. D. Breach and F. W. Wulff, "Intermetallic Growth in Gold Ball Bonds Aged at 175 °C: Comparison between Two 4N Wires of Different Chemistry," *Gold Bulletin,* vol. 42, no. 2, pp. 92–104, 2009.
[27] M. A. Bahi, H. Fremont, J.-P. Landesman, A. Gentil and P. Lecuyer, "A New Methodology for the Identification of Ball Bond Degradation During High-temperature Aging Tests on Devices in Standard Plastic Packages," *Microelectronics Reliability,* vol. 49, pp. 1273–1277, 2009.
[28] R. D. Maldonado, A. I. Oliva and J. E. Corona, "Annealing Effects on the Electrical Resistivity of AuAl thin Films Alloys," *Materials Chemistry and Physics,* vol. 116, pp. 453–457, 2009.
[29] M. Bahi, P. Lecuyer, A. Gentil, H. Fremont and J.-P. Landesman, "Package Delamination as Indicator of Ball Bond Lift: New Diagnostic Methodology," in *Thermal, Mechanical and Multi-Physics simulation and Experiments in Microelectronics and Microsystem, 2009, EuroSimE 2009, 10th International Conference on*, 26–29 April 2009.
[30] J.-W. Shin, M.-S. Song, Y.-S. Park, Y.-M. Kwon, J.-T. Moon, J.-S. Cho, K.-A. Yoo, K.-Y. Byun, C.-H. Joh, E.-H. Do and K.-W. Paik, "Analysis on Interfacial Failures of Ultra-fine Pitch Wire with Low Inter-metallic Coverage," in *Electronic Packaging Conference, 2009. EPTC '09. 11th*, 9–11 December 2009, pp. 942–946.
[31] H. Xu, C. Liu, V. Silberschmidt, Z. Chen and M. Sivakumar, "TEM Interfacial Characteristics of Thermosonic Gold Wire Bonding on Aluminium Metallization," in *Electronics Packaging Technology Conference, 2009. EPTC '09. 11th*, 9–11 Dec. 2009.
[32] C. W. Tok and C. Effie, "Ultra-fine-wire Applications and Challenges in 30um Process," in *2009 11th Electronics Packaging Technology Conference 9-11 December 2009*, Singapore, 2009.

[33] M. McCracken, H. J. Kim, M. Mayer, J. Persic, J. S. Hwang and J.-T. Moon, "Assessing Au-Al Wire Bond Reliability using Integrated Stress Sensors," in *Thermal and Thermomechanical Phenomena in Electronic Systems (ITherm), 2010 12th IEEE Intersociety Conference on*, 2–5 June 2010.

[34] M. Krause, B. Maïrz, C. Dresbach and M. Petzold, "Electron Backscatter Diffraction Microstructure Investigations of Electronic Materials down to the Nanoscale," in *Electronic System-Integration Technology Conference (ESTC), 2010 3rd*, 13–16 September 2010.

[35] M. Krause, M. Petzold, B. März and S. Bennemann, "*High Resolution Analysis of Intermetallic Compounds in Microelectronic Interconnects using Electron Backscatter Diffraction and Transmission Electron Microscopy*," in *Electronic Components and Technology Conference ECTC*, 2010.

[36] B. März, S. Scheibe, A. Graff and M. Petzold "Growth Behaviour of Gold-Aluminum Intermetallic phases (IMP) in Temperature Aged Ball Bonds observed by Electron Backscatter Diffraction," in *Electronic System-Integration Technology Conference (ESTC) 2010 3rd*, 13–16 September 2010.

[37] H. Xu, C. Liu, V. V. Silberschmidt, S. S. Pramana, T. J. White, Z. Chen, M. Sivakumar and V. L. Acoff, "A Micromechanism Study of Thermosonic Gold Wire Bonding on Aluminum Pad," *Journal of Applied Physics,* vol. 108, p. 113517-1-8, 2010.

[38] Y. Jiang, R. Sun, S. Wang, D. Min and W. Chen, "Study of a Practicable Wire Bonding Method for Applying Copper Wire Bond to Large-scale Integrated Circuits," in *Electronic Components and Technology Conference (ECTC), 2010 Proceedings 60th.*, 1–4 June 2010.

[39] Y. Jiang, R. Sun, Y. Yu, Z. Wang and W. Chen, "Formation and Growth of Intermetallic Compounds in Cu–Au and Au–Al Systems for Copper on Gold Bonding," *IEEE Transactions on Electronics Packaging Manufacturing,* vol. 33, no. 3 (July), pp. 228–235, 2010.

[40] J. Beleran, T. H. Hong, O. P. Wilson, X. R. Zhang, G. L. Amurao, G. Mehta and W. Zhu, "Solutions for 45/40nm ELK Device Integration into Flip Chip and Wire Bond Packaging," in *2010 Proceedings 60th Electronic Components and Technology Conference (ECTC) 1-4 June 2010*, Las Vegas NV, 2010.

[41] M. O and M. Kajihara, "Kinetics of Solid-State Reactive Diffusion between Au and Al," *Materials Transactions,* vol. 52, no. 4, pp. 677–684, 2011.

[42] C. D. Breach and T. K. Lee, "Conjecture on the Chemical Stability and Corrosion Resistance of Cu-Al and Au-Al Intermetallics in Ball Bonds," in *Electronic Packaging Technology and High Density Packaging (ICEPT-HDP), 2011 12th International Conference on*, 8–11 August 2011.

[43] B. Yao, P. Lai, J. Liu and X. Liang, "The Effect of Thermal Stress on High Density Packaging Integrated Circuits," in *Advanced Packaging Materials (APM) 2011 International Symposium on*, 25–28 October 2011.

[44] B. Yao and P. Lai, "Accelerated Reliability Evaluation for High Density Packaging Integrated Circuits," in *Quality, Reliability, Risk, Maintenance, and Safety Engineering (ICQR2MSE), 2011 International Conference on*, 17–19 June 2011.

[45] D. Lv and G. Yang, "Gold Bonding Reliability in High Density FPGA Devices," in *2011 International Conference on Electronics and Optoelectronics (ICEOE)*, 29–31 July 2011.

[46] H. Xu, V. Acoff, C. Liu, V. Silberschmidt and Z. Chen, "Void Growth in Thermosonic Copper/Gold Wire Bonding on Aluminum Pads," in *Electronic Components and Technology Conference (ECTC), 2011 IEEE 61st.*, May 31 2011–June 3, 2011.

[47] J. Delucca, J. Osenbach and F. Baiocchi, "Observations of IMC Formation for Au Wire Bonds to Al Pads," *Journal of Electronic Materials,* vol. 41, no. 4, pp. 748–756, 2012.

[48] R. Huang, Y. Y. Tan, J. Walter, H. Pape, X. Fan and H. Koerner, "Simulation of Diffusion Controlled Intermetallic Formation of Au/Al Interface," in *Thermal, Mechanical and Multi-Physics Simulation and Experiments in Microelectronics and Microsystems (EuroSimE) 2012 13th International Conference on*, 16–18 April 2012.

[49] C.D. Breach and T.K. Lee, "Shear Strength and Failure Modes of As-Bonded Gold and Copper Ball Bonds on Aluminum Metallization," Journal of Electronic Materials, vol. 41, no. 7, pp. 2018–2028, 2012.

[50] S. S. Ahmad and S. C. Smith, "Au/Al Wire Bond Interface Resistance Degradation Rate Simulations," *IEEE Transactions on Device and Materials Reliability,* vol. 19, no. 4, pp. 774–781, December 2019.

13 Assessment of Au-Al Bond Interface Integrity

This chapter briefly lists some methods commonly used to evaluate the gold-aluminum interface.

In the semiconductor packaging studies and evaluations, the quality, integrity, health, and degradation of the Au-Al bond is measured in a number of ways.

Mechanically, it is measured by pulling the wire or by shearing the Au-Al joint (Figure 13.1). The results of the wire pull test are given in the form of the pull strength in gram-force (gf) and the corresponding failure modes (Figure 13.2).

In a wire pull test, when the wire fails in the mid-span, the only thing the pull test tells is that the gold ball bond to the aluminum bonding pad on the chip and to the bonding pad on the substrate or lead frame is stronger than the wire itself. As-made unstressed bonds made using an optimized ball bonding process are expected to fail at wire mid-span as the gold ball bond area to the bonding pad on the chip is two to ten times larger than the cross-sectional area of the wire itself. Failure at other locations may only occur if the process is not set up properly.

If the failure occurs at the neck of the ball bond to the pad chip or at the neck of the stitch bond to the substrate or lead frame, it generally offers little insight into the condition of the Au-Al bond itself. Instead, it simply indicates a weakness at the neck of the ball bond or at the neck of the stitch bond. Reasons for such failures have been studied extensively over the years [1].

The only time the pull test provides information about the Au-Al interface is when the Au-Al bond lifts off the aluminum bonding pad on the chip. What this means is that either it was an un-optimized bond to begin with, the Au-Al bond had deteriorated to the extent that its interfacial area or strength had been reduced significantly, or there was a fracture at the Au-Al bond interface. This can be misleading [2]. Additionally, mechanical testing is destructive.

The ball shear test (Figure 13.3) provides insight into the bond interface condition. It can cause failure through Au or Al metals, can rip silicon out from under the Al, or shear through the intermetallic, eliciting the weakest part of the structure (Figure 13.4). This test again is destructive, though it is more useful than the wire pull test in determining the condition of the Au-Al interface. Shear value, measured in gf, is dependent upon the gold ball diameter or the Au-Al contact interface area. Pictures of different failure modes are shown in Figure 13.4.

DOI: 10.1201/9781003619178-13

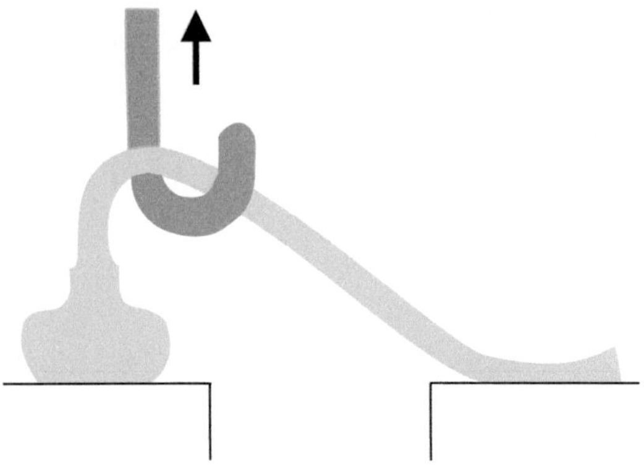

FIGURE 13.1 Wire pull method to test the condition of a wire bond.

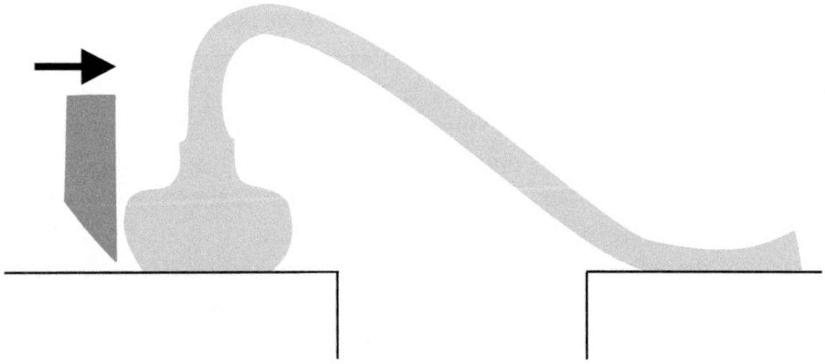

FIGURE 13.2 Wire pull failure modes.

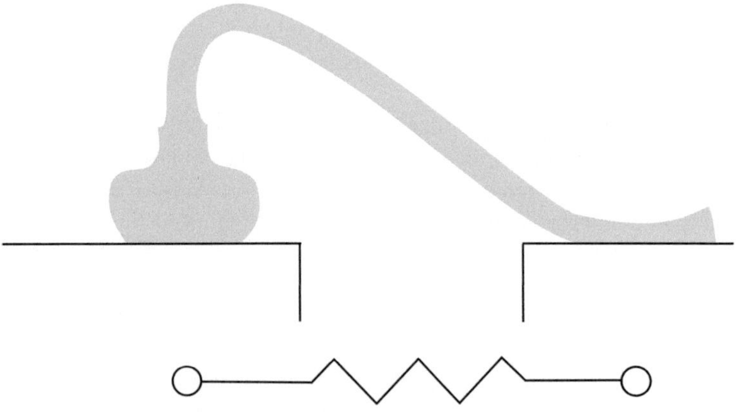

FIGURE 13.3 Ball shear testing method for Au-Al interface integrity.

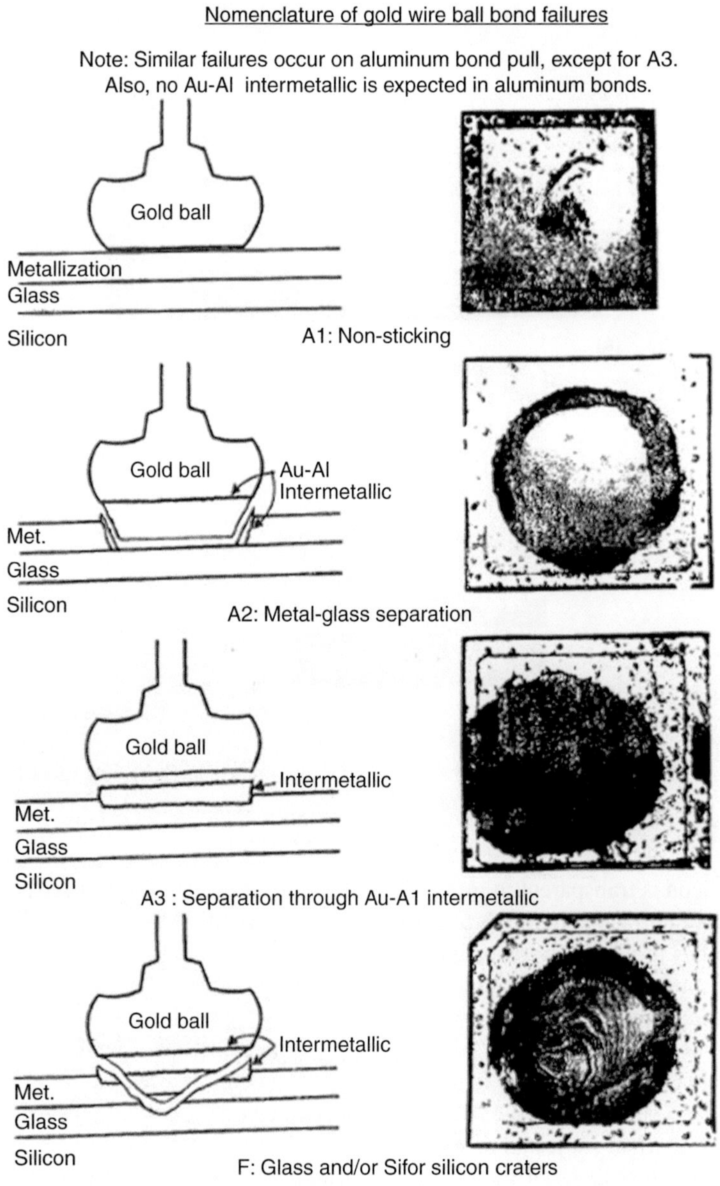

FIGURE 13.4 Ball bond failure modes [3].

The Au-Al bond interface is also often sectioned and polished to observe the metallurgy of the interface and its condition, e.g., Figure 13.5. This method again is destructive. It also is time intensive and cannot be used in-situ. It is good for metallurgical studies, but is not practical for general evaluations.

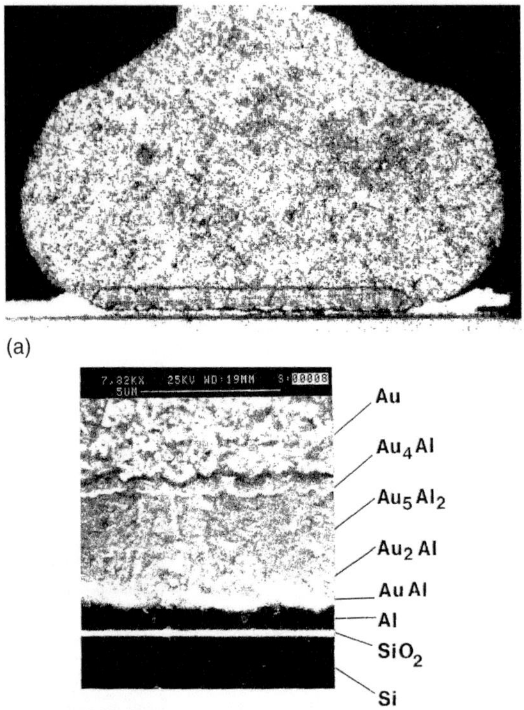

FIGURE 13.5 A cross section of a gold-aluminum interface (left) and identification of different intermetallic phases (right) [4].

As silicon is transparent to infrared (I) IR microscopy can also be used to observe the Au-Al interface (Figure 13.6 and Figure 13.7). As the packaging material under the chip has to be removed for interface observation, the process is destructive and time consuming, and is not suitable for general evaluations.

Electrically, the health of the joint is looked at by measuring resistance changes in the interconnect (Figure 13.8). As the Au-Al interface degrades, the resistance of the bond increases. This method is non-destructive and can be used for monitoring the integrity and condition of the Au-Al bond instantaneously, continuously through its life cycle, or when subjected to stress testing.

When the resistance of the bond is measured (e.g., Figure 13.9), it includes the resistance of the bonding post on the chip and that on the external carrier, the resistance of the wire, the resistance of the gold and stitch bonds and the resistance of the Au-Al interface. As the resistance of the Au-Al interface deteriorates, the resistance in the other parts of the bond remains practically unchanged, especially when compared to the large changes occurring at the interface. On degradation of the interface, the resistance at first "changes" very minutely while the intermetallic compounds are formed. Larger changes to resistance start taking place when Kirkendall voids start to show up. These changes accelerate in the presence of halides.

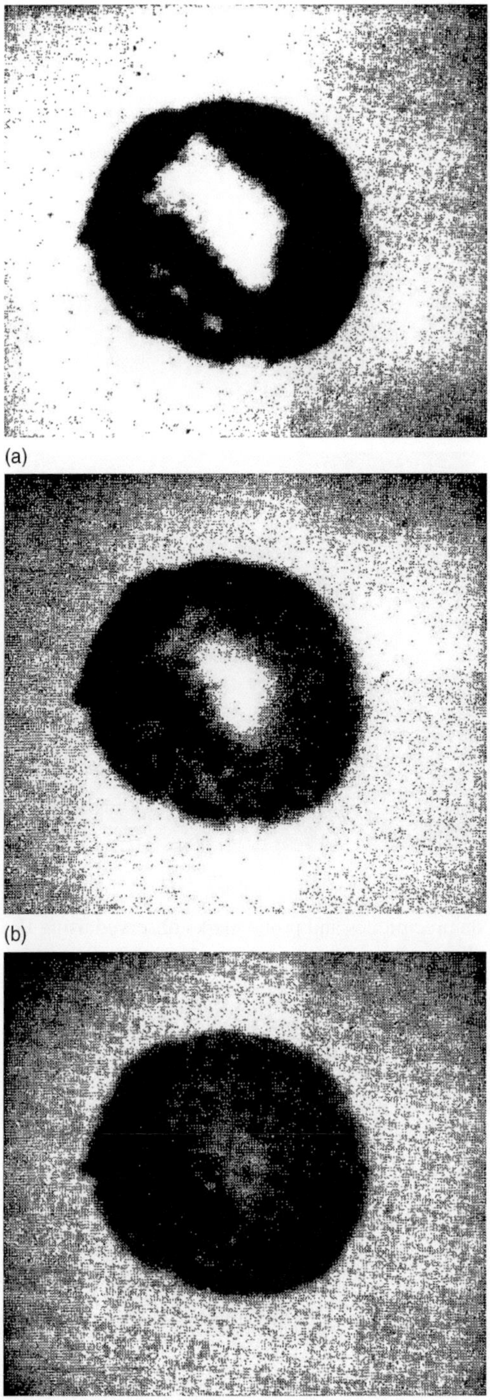

FIGURE 13.6 Intermetallic proliferation observed through IR [4].

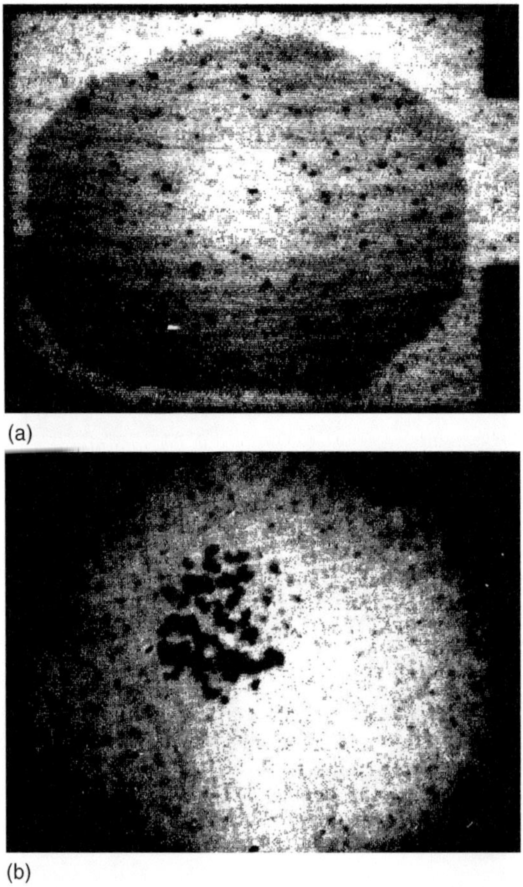

FIGURE 13.7 Silicon precipitates and probe marks observed using IR [4].

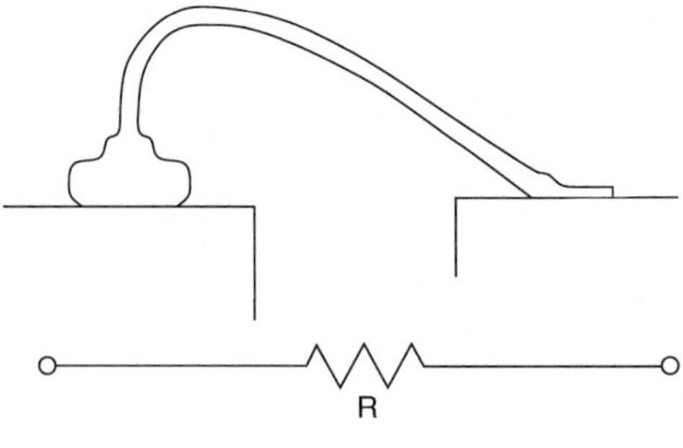

FIGURE 13.8 Gold ball bond resistance.

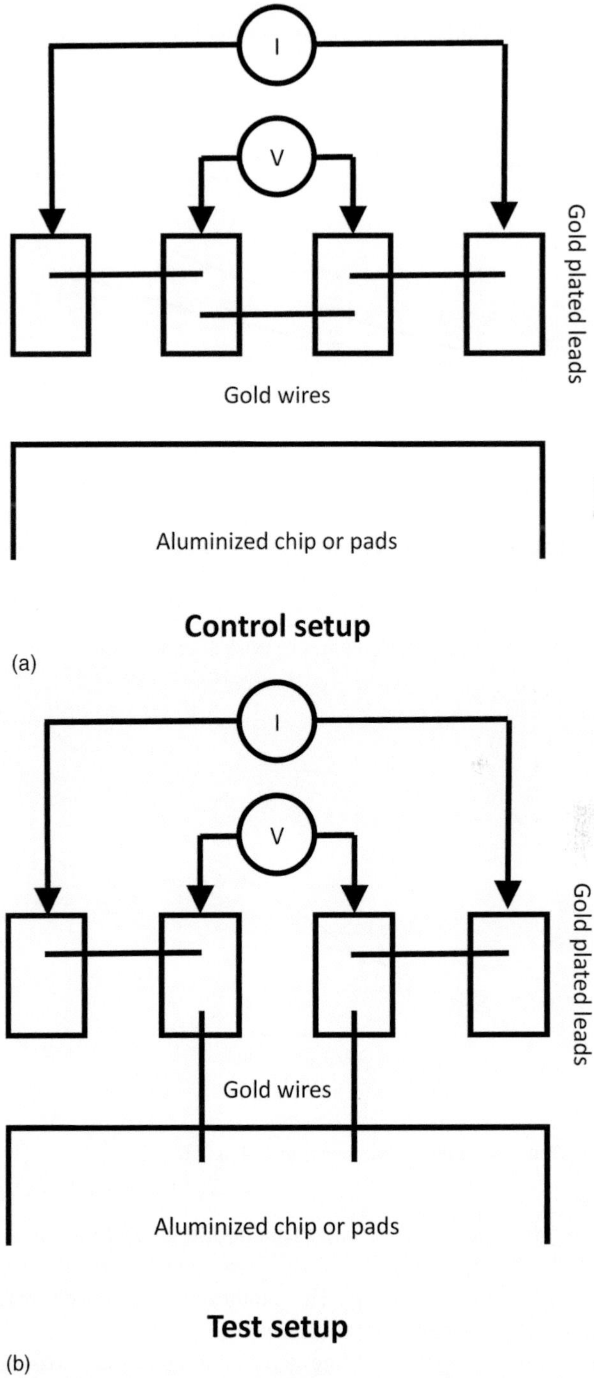

FIGURE 13.9 Resistance measurement arrangements for control and test structures.

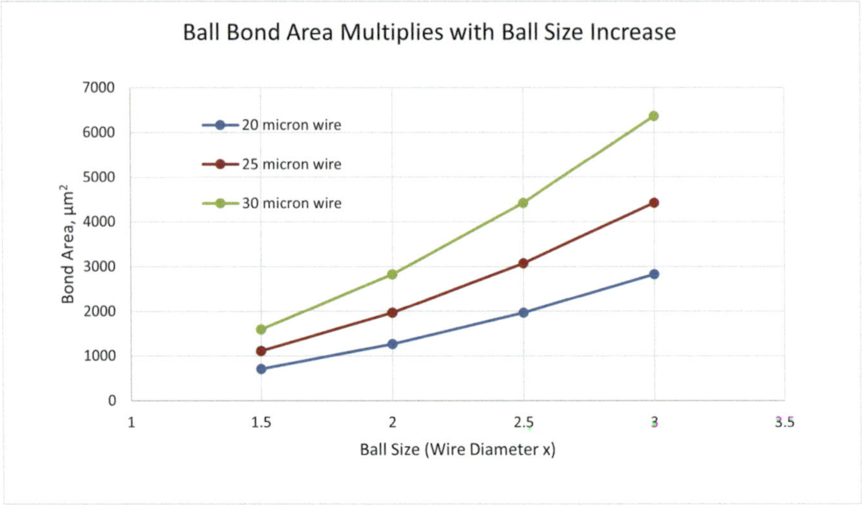

FIGURE 13.10 Ball bond interconnect area increases with increasing ball size [3].

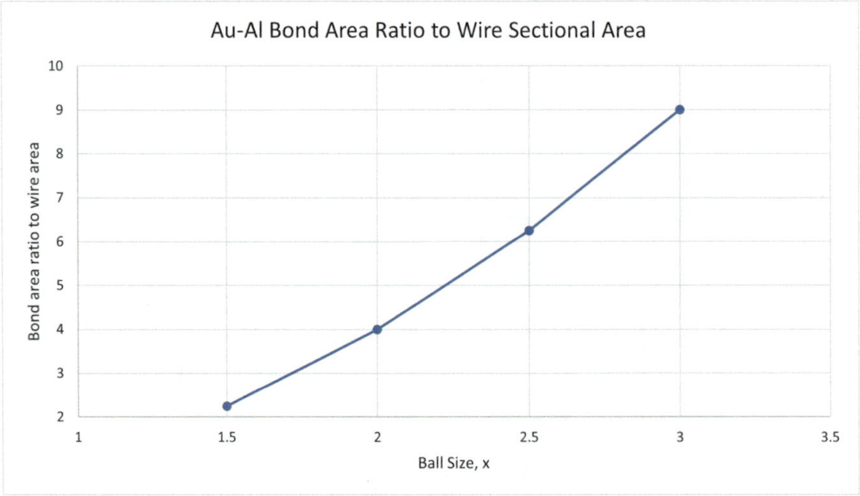

FIGURE 13.11 Bond area ratio to wire sectional area [3].

As resistance provides the opportunity for continuous and in-situ measurement of the condition of the Au-Al interface, it is the favored choice for the evaluation of the quality and reliability of Au-Al bonds. This method has been used by researchers for decades due to its efficiency and effectiveness.

As the gold ball is larger than the wire diameter, the cross-sectional gold-to-aluminum bond area is much larger than the cross-sectional area of the gold wire itself. Unless the bond is weak to begin with, as degradation at the gold-aluminum interface reduces the contact area, failure will continue to take place in the wire until the Au-Al contact area is reduced sufficiently to become weaker than the wire itself.

As shown in Figure 13.10 and Figure 13.11, the Au-Al contact area is much larger than the gold wire sectional area; therefore, wire pull testing can show a failure at the gold-aluminum interface only after a large reduction in the Au-Al interface area.

REFERENCES

[1] G. Harman, *Wire Bonding in Microelectronics*, McGraw-Hill, 2010.

[2] R. G. Oswald, W. R. R. D. Miranda and C. W. White, "Changes in Strength and Resistance of Aluminum to Gold Ultrasonic Bonds after Temperature, Electrical, and Environmental Stress," in *Proceedings 1975 International Microelectronics Symposium,* pp. 324–339, 1975.

[3] S. S. Ahmad, "Characterizing the Aging-driven Degradation Rate of Electrical Contact Resistance and Mechanical Integrity of Plastic-encapsulated Au/Al Wire Bonds," NDSU, Fargo ND, December 2019.

[4] P. K. Footner, B. P. Richards, C. E. Stephens and C. T. Amos, "A Study of Gold Ball Bond Intermetallic Formation in PEDs using Infra-Red Microscopy," in *24th Annual Reliability Physics Symposium*, April 1986.

14 Summary

Another important transition that affected gold wire reliability to aluminum bond pads on integrated circuits was the bonding method itself. The early days used "thermocompression" bonds in which the gold ball was pressed to the silicon integrated circuit chip at roughly 300 °C. These bonds were substantially overprocessed at manufacturing time, so the reliability under bake stress was poor. The industry gradually morphed to using "thermosonic" bonding, in which the gold ball was still pressed against the aluminum pad, **but** ultrasonic energy inserted the ball enabled using a much smaller chip temperature, typically 150 °C. Thus, gold-aluminum bond reliability was substantially improved.

The next big change was the realization that some gold wire dopants (impurities) improved bond reliability. Thus, the norm is now to use gold wire with ~1 % impurities rather than perfectly pure gold (5 9's). Ironically, gold wires are now largely supplanted by copper wires for cost reduction. Copper wire requires major changes during bonding due to copper oxidation, so the copper wire is usually coated with palladium. However, coated copper wires are superior in bake performance and during exposure to humidity.

DOI: 10.1201/9781003619178-14

15 Glossary

2N	(2 nines, i.e., 99%)
3N	(3 nines, 99.9%)
4N	(99.99%, 100 ppm) purity
ANSYS	Analysis System — an analysis software manufacturer.
APM	advanced packaging materials
ASTM	American Society for Testing and Materials
BGA	Ball Grid Array — an n-by-m array of bonds, typically spaced at a coarser pitch than peripheral-only connections, but offering greater areal efficiency.
BP	Biphenyl-based molding compound
BPSG	borophosphosilicate glass
BPT	ball pull test
C4	Controlled Collapse Chip Connection
CEN resins	Cresolic epoxy novolac resins
CerDIP	ceramic dual-in-line package
COC	Chip-On-Chip, in which one chip is flip-chip bonded to the face of another chip.
COF	Chip-On-Flex/Film — when a chip is attached to a flexible substrate such as a tape or small flexible board.
COG	Copper on gold.
COG	chip-on-glass.
CTE	coefficient of thermal expansion
CVD	chemical vapor deposition
DIP	dual in-line package
EBSD	electron backscatter diffraction
EDS	energy-dispersive X-ray spectroscopy used with an electron microscope to determine the elemental composition
EHOF	effective heat of formation
EMC	Epoxy-based Molding Compound — a thermosetting resin with good stability and low shrinkage.
ESD	electrostatic discharge
eV	electron volt

DOI: 10.1201/9781003619178-15

FA	failure analysis
FC	flip-chip
FCC	Face-Centered Cubic — a common crystal structure for noble metals such as Au, Ag and other metals such as Cu and Al.
FEM	Finite Element Modeling — a technique that uses continuity of differential equations over interfaces between zones with different mechanical properties or irregular structure, hence one can numerically estimate stress and strain for configurations that cannot be solved in closed form.
FIB	Focused Ion Beam — an analysis technique as in SIMS (Secondary Ion Mass Spectrometry) or as a molecularly based tool to remove materials for failure analysis (dig a trench to permit compositional analysis). Also used to cut electrical traces or to add materials to create an electrical connection, used for design debugging and failure analysis.
FIMV	force current measure voltage — used to measure resistance.
FPGA	fine-pitch grid-array
FR	flame retardant — material component in molded electronic assemblies to stop fire.
Freezing Point Method	freezing point of a solution is lower than that of the solvent after the addition of a solute. The molecular weight of the solute can be determined by using the freezing point depression and the molality of the solution.
HAST	Highly Accelerated Stress Testing — a very high humidity and temperature stress, simulating accelerated "jungle" conditions.
HR	High Reliability
HRTEM	high-resolution transmission electron microscope
HS	High Security
HTS	High Temperature Storage to simulate shelf-life estimation.
HTST	high temperature storage test
IC	integrated circuit
IMC	InterMetallic Compound —a stoichiometric combination of two or more metals, exhibiting mechanical and electrical properties that are significantly different from those of its individual metal components.
IMP	intermetallic phases
IP	intermetallic phase
IPC-MS	Inductively Plasma-Coupled Mass Spectrometer — a very sensitive method to produce excited ions from a solid for compositional analysis.
ICP-OES	Inductively Coupled Plasma-Optical Emission Spectrometry — similar to IPS-MS but detection is optical rather than mass spectrometer.
IR	infrared

LCD	liquid crystal display
LSI	Large-scale integration
MC	molding compound
MCM	Multi Chip Module — used when disparate types of chips cannot be easily produced as a single chip and/or when a large chip size would drive a very small yield.
MOSFET	metal-oxide-semiconductor field effect transistor
MIL-STD	Military Standard
MTC	Monitored Temperature Cycle
OCN	OrthoCreosol-Novolac — molding compound.
P-DIP	plastic dual in-line package
PED	plastic encapsulated device
PEM	Plastic Encapsulated Microcircuit, also known as, a plastic IC (an IC packaged in plastic housing) rather than hermetic (an IC sealed hermetically in non-moisture absorbing materials)
PIP	package in package
PSG	phosphorus-doped silicon oxide — phosphosilicate glass.
RDS(on)	drain-to-source-(on) resistance
SAB	Surface Activated Bonding — atomically clean and activated surfaces of semiconductors, metals and dielectrics can form direct bonds between them on contact at room temperature.
SEM	scanning electron microscope
SIMS	Secondary Ion Mass Spectrometry
SIP	system in package
TAB	Tape Automated Bonding
TBBA/TBBPA	tetrabromobisphenol A
TC bond	thermocompression bond
TEM	transmission electron microscope
T_g	glass transition temperature
T/S	thermal shock
UBM	under-bump metallurgy
UFP	ultra-fine pitch
UFW	ultrafine-pitch wire
UHV	Ultra High Vacuum
VLSI	very large scale integration
WB	Wire Bond — a fine wire (Au, Al or Cu) connecting a pad on the semiconductor chip (Si, GaAs, etc.) to a substrate or package, typically at a substantially different pitch, so wire bonds are both a connection technology and a pitch transformation device.

Index

A

Air, effect of, *see* Ambient
Aluminum
  contact spiking, 59
  effect of over- and under-etch of Al on pad, 61
  pure Al *vs.* Si-doped Al, 54
  reasons for using aluminum, 13, 16
Ambient
  ambient ball/gold bonding, 72–73, 94
  effect of ambient, 10, 13, 25, 26, 48
antimony (Sb)
  antimony oxide as a flame retardant, 63, 66
  effect of, 66, 93, 97
  in wire, 35
Aqueous extraction, *see* Water
Argon, effect of, 47, 48
Arrhenius curve, 24, 47
Arsenic (As), effect of, 91, 93, 96
Auger electron spectroscopy (AES), 49–50, 71

B

Beryllium, effect of addition to gold wire, 18, 22, 50, 92
Biphenyl epoxy, 76, 95
Bisphenol, 63, 97
  tetrabromobisphenol A, 63
Bismuth, 41
Black death/black plague, 10, 14, 16, 17, 21, 33, 41
  Si is not required, 33
Black plague, *see* Black death
Bond
  geometry, 32, 53, 58
    recommendation, 38
    and testing, 44
    and modeling, 62
    and probe damage, 81
  see also Thermocompression; Thermosoinic/ultrasonic
Boron, impurity, 33
Brass, Cu Zn interdiffusion, 5
Bromine,brominated resin as flame retardant, effect of, 48, 55, 58–60, 63, 69, 76, 96, 97

C

Calcium, effect of, 92
Capillary
  effect of metal capillary, 50
  effect of temperature of, 36
Chlorine effect of, 53, 55, 60, 107
  chloride effect of, 49, 87
Cobalt, impurity, 33
Contamination effect of, 52–55, 69–71, 92, 97
  chlorine, 60
  initial voids, 84
  measurement, 74
Controlled Collapse Chip Connection (C4), 70
Copper-aluminum, 11, 81, 83, 84
Copper and beryllium, 50
Copper bonding, effect of Al thickness, 93
Copper, effect of, 35, 41, 50, 64, 81, 91, 92, 95, 96, 107, 109
Copper in bonding, 93, 94, 97, 98, 100, 104, 106, 110, 124
Copper resistivity compared to $AuAl_2$, 13
Copper zinc interdiffusion, 5
Cresolic epoxy novolac resins, 49, 62, 63, 68, 76, 95, 97
Curve tracerto identify opens, 14

D

Delamination monitoring, 102
DGETA [diglyceryl ether of tetrahydro anhydrohexitol], effect of, 66
Diffusion bonding, solid phase, 72–75, 82, 98, 107
  modeling, 54

E

EDS/EDX, use of, 86, 88, 99
effective heat of formation rule, use of, 67–68, 107
electron intermetallic compounds, 61
ellipsometry, 74
epoxy, effect of, 52–55, 67
  brominated epoxy, 60
  epoxy die boning, effect of, 67, 69
ESD structures, 81

F

Fe, effect of, 33, 92
FIB, *see* Focused ion beam
Fick's Law and Fickian behavior, 10, 32, 39, 58, 63–64, 76, 86, 91–92, 94, 96, 100–101, 105, 107, 110
Flame retardants and effect of, 48, 49, 55, 63, 66, 69, 96, 97; *see also* Antimony oxide; Bromine, brominated resin

Fluorine and its effects, 3, 53
Focused ion beam (FIB), use of, 50, 93
Forming gas (N_2 + 10 % H_2), 26

G

Galvanic corrosion/effect, 34, 66
Germanium in gold wire/effect of/role, 3, 8, 61
  bonds to Ge, 11
Glass transition temperature, effect of, 49
Gold
  dopants/impurities, 8, 35, 50, 76, 92–94, 101, 107, 124
  effect of addition of Be, 18, 22, 50, 92
  germanium in gold wire, 8, 61
  impurities (*see* Dopants)
  outgassing blowholes in gold wire, 35
  properties, 35
  reflectivity, 37
  teardrop ball, 38
  using gold, 11, 12
  wire bond loop height and bond flexure, 38
  wire manufacture, 35

H

Helium, effect of, 48
Hermetic packages *vs.* plastic packages, 48, 87
Hydrogen
  effect of, 47, 48
  effect of flame-off, 35, 50

I

ICP-MS/OES, 96, 97
Infrared microscopy, 60, 118, 119, 120
Infrared thermal measurements, 46
Intermetallic compounds, *see* Purple plague
  valency, electron, interstitial, 61
  Al_3Au_8 and not Al_2Au_5, 62
  highlighting, 98, 99
Intermittent opens/failures, 11, 21, 34, 40, 59
Interstitial intermetallic compounds, 61
Ion beam micro-texturing, 50

K

Kirkendall effect and void formation, 37, 40, 41, 46, 47, 71, 74, 75, 82, 83, 86, 94, 98, 100, 101
  accompanied by electromigration, 38

L

Lead (Pb) in wire, 35
  role of, 93

M

Magnesium (Mg) in wire, 35, 38, 44
Meta-bromo phenol, 63
Moiety, 63
Moisture, *see* Water
Multi-aromatic resin (Green resin), 96

N

Nickel, 33, 54, 70
Nitrogen
  bonding in, 94
  effect of, 48, 70

O

OCN, *see* Cresolic
Ortho cresol novolac OCN, *see* Cresolic

P

Palladium
  effect of, 41, 86, 91–93, 99, 100, 102, 107
  coated copper wire, 124
Parylene application, 40
Phases, Au/Al and other phases, *see* Purple plague
Phase diagrams, 87
Phosphosilicate glass, effect of PSG layer, 59
Plastic packages *vs.* hermetic packages, 48
Platinum, effect of, 8, 41, 91, 107
  TiPt barrier layer, 67
Pluck, bond pluck test, 56
Polyimide, effect of, 70
Power pulses, effect of, 100
Probe marks, effect of, 81, 84
  infrared observation, 120
Pull (bond wire pull), 99, 101, 115, 116
  data can be misleading, 40, 44
Pull test, non-destructive, 53, 56
Purple plague and other Au/Al phases, intermetallic compounds
  activation energy, 11–12, 14, 23–24, 28, 32–33, 39, 49, 50, 53–55, 59, 64, 68, 71, 74, 76, 84, 88, 91–92, 95, 97, 106
  amine and phosphorus initiators, effect of, 68
  Au_8Al_3 and Au_5Al_2, 105
  crystallographic data, 105
  exponential growth, 58
  flexure fatigue, susceptibility to, 39
  missile electronics, 45
  phosphorus initiators, effect of, 68
  properties, 12, 13
  rubber modifiers, effect of, 68
  strain/stress caused by phases, 39, 49
  substrate, effect of, 48

volume changes, 3, 14, 20, 26, 40, 59, 82–83, 85, 96, 109–110
a third layer of intermetallic grows between Au_4Al and Au_5Al_2, 83
use of piezoresistive integrated CMOS microsensors, 103–104

R

RDS(on), drain- to- source on- resistance, 97, 98
Reflectivity, 11–12, 37
and reflectance theory, 74
Repairable bonding process, 72–73
Residual resistance ratio
definition, 13
Resistance
deduced from ESD protection diodes, 101
linear resistance drift, 68
resistance and bond integrity, 24, 44, 49, 54, 55, 76
resistance change in intermetallic compounds at joint, 22, 23, 32, 33, 63, 94, 110
testing, 44, 103, 104, 108, 116, 118, 120–2
Resistivity, 40, 71
and changes in grain size, 102

S

SAB, *see* Surface activated bonding
SAC (Sn-Ag-Cu) solder, 104
Sapphire, effect as substrate, 48
Scanning electron microscope, use of, 47, 48, 67, 72, 86, 88, 93, 99, 102
electron backscatter diffraction (EBSD), 39, 50, 104, 105
Self-healing, 34
Shear
ball shear induced cratering, 62, 64–66
shear failure mechanisms, 55, 64
strength, circumference *vs.* area, 55
testing/results, 44, 56, 63, 81, 94, 100, 101, 110, 115, 116
Silicon
alloying mechanism, 8–9
Al-Si-Cu pad, 66, 67
catalyst, as, 10, 16, 24, 41
cratering of, 62, 64–66, 115, 117
from adhesives/silica, 52
infrared for examination, 60
in gold wire, 35, 44
Mg *vs.* Si in Au wire, 38, 44
Ni-Si, 54
nodule formation by precipitation and acting as points of high stress, 59
refractory plugs to isolate Si from Al, 59
oxidation, effect of, 13
residue on, 61
silicon as substrate effect of, 48, 54–55, 63, 71, 88
ternary compound with Al, 20, 41
why Al on Si, 16
Silicone rubber, effect of, 49
Silver, 1, 7, 8, 11, 13, 26, 38, 87
gold-silver interface, 98
in wire, 35
x-ray spectroscopy, 24–25
SIMS use of, 52, 72
Sodium, 52
Sodium chloride, effect of, 34
substrate, 39
Solubility, 3–4, 9, 86
chloride, 34
STEM, 91
Surface activated bonding (SAB), 82, 83, 88, 94
and soldering, 94

T

Tape automated bonding (TAB), 49, 69–70
TEM, 39, 64, 70, 71, 93, 98, 102–104
Tetrabromobisphenol A, *see* Bisphenol
Thallium accumulation, 49, 50
Thermocompression bonding, 7, 10, 12, 21–22, 25, 35, 36, 38, 41, 50, 69, 124
Thermosonic bond, 84, 86–88, 92–93, 103–105, 109, 124
Titanium, 44, 82
TiPt barrier layer, 64, 67
Transmission electron microscope (TEM) use of, 64, 70, 98, 104
identification of phases using, 39

U

Ultrasonic bonding, 38, 63, 98, 103, 106
gold to gold, 45–46
non-concentricity for 2N wire, 103

V

Vacuum effect of, 70, 88, 94, 95, 102
Valency intermetallic compounds, 61
Voiding
annular/peripheral/torus voiding at ball bond interface, 40, 55, 63, 64, 84, 86, 109
initial voiding by probe marks, 81, 84
infrared observation, 120

W

Water
- aqueous extraction, use of, 49
- corrosion/degradation, 66, 69–70, 76, 87, 95, 107
- role in dendritic growth, 13
- role in diffusion, 69
- trapped insoluble, effect of, on grain boundaries, 50

X

X-ray diffraction, use of, 3, 7–9, 24–25, 39, 49–50, 71
- backscattering, 39, 50
- florescence, 49
- photoelectron spectroscopy, 71, 99

Z

Zinc copper interdiffusion, 5

For Product Safety Concerns and Information please contact our
EU representative GPSR@taylorandfrancis.com Taylor & Francis
Verlag GmbH, Kaufingerstraße 24, 80331 München, Germany